The Planet in 2050

In 2050, the billions of people living on Earth have found a way to manage the planetary system effectively. Everyone has access to adequate food, shelter and clean water. Human health is no longer considered outside of the health of the ecosystems in which people live. Ecological awareness is an integral part of education. People respond effectively to social and environmental hazards, and societies care for the most vulnerable amongst them. The economy, too, has shifted. Carbon dioxide management is under control, and energy efficiency is the norm. The remaining rainforests have been preserved. Coral reefs are recovering. Fish stocks are thriving. Is any of this really possible? How can our complex social and economic systems interact with a complex planetary system undergoing rapid change to create a future we all want?

This book is a contextualized collation of ideas articulated by the 50 participants of the Planet 2050 workshop held in Lund in October 2008, as part of *The Planet in 2050*, an interdisciplinary Fast Track Initiative of the International Geosphere-Biosphere Programme. Participants were selected from academia and the sustainability practice community to give a wide-ranging, multi-cultural, transdisciplinary set of perspectives.

This collection explores four broad sectoral themes: energy and technologies; development, economies and culture; environment; and land use change. By doing so, this book emphasizes the importance of a social dialogue on our collective future, and our responsibility to the Earth. It makes strong statements about what needs to happen to the global economy for a sustainable future and documents a new kind of scholarly discussion, engaging people from diverse knowledge communities in a spirit of exploration and reflexivity. The book provides a focus for dialogue and further study for postgraduates and researchers interested in global change as a multi-faceted, socio-environmental phenomenon, and as the book is written in an accessible scholarly style, assuming no prior specialist knowledge, it is also suitable for those involved in sustainability initiatives and policy.

Jill Jäger is an independent scholar and editor of several books on sustainability and policy. **Sarah Cornell** is a research fellow at the University of Bristol, UK, and is also responsible for the scientific coordination of the UK Natural Environment Research Council's research programme on Earth system science, QUEST (*Quantifying and Understanding the Earth System*).

Routledge studies in ecological economics

1 **Sustainability Networks**
Cognitive tools for expert collaboration in social-ecological systems
Janne Hukkinen

2 **Drivers of Environmental Change in Uplands**
Aletta Bonn, Tim Allot, Klaus Hubaceck and Jon Stewart

3 **Resilience, Reciprocity and Ecological Economics**
Northwest coast sustainability
Ronald L. Trosper

4 **Environment and Employment**
A reconciliation
Philip Lawn

5 **Philosophical Basics of Ecology and Economy**
Malte Faber and Reiner Manstetten

6 **Carbon Responsibility and Embodied Emissions**
Theory and measurement
João F.D. Rodrigues, Alexandra P.S. Marques and Tiago M.D. Domingos

7 **Environmental Social Accounting Matrices**
Theory and applications
Pablo Martínez de Anguita and John E. Wagner

8 **Greening the Economy**
Integrating economics and ecology to make effective change
Bob Williams

9 **Sustainable Development**
Capabilities, needs, and well-being
Edited by Felix Rauschmayer, Ines Omann and Johannes Frühmann

10 **The Planet in 2050**
The Lund discourse of the future
Edited by Jill Jäger and Sarah Cornell

The Planet in 2050

The Lund discourse of the future

Edited by Jill Jäger and Sarah Cornell

Routledge
Taylor & Francis Group
LONDON AND NEW YORK

First published 2011
by Routledge
2 Park Square, Milton Park, Abingdon, Oxon OX14 4RN

Simultaneously published in the USA and Canada
by Routledge
711 Third Avenue, New York, NY 10017, USA

Routledge is an imprint of the Taylor & Francis Group, an informa business

Typeset in Times by Wearset Ltd, Boldon, Tyne and Wear

British Library Cataloguing in Publication Data
A catalogue record for this book is available from the British Library

Library of Congress Cataloging in Publication Data
The planet in 2050: the Lund discourse of the future/edited by Jill Jäger and Sarah Cornell.
p. cm.
Collation of ideas from Planet 2050 workshop held in Lund and Malmö, Sweden, Oct. 27 to 31, 2008, as part of Planet in 2050, interdisciplinary fast track initiative of International Geosphere-Biosphere Programme.
"Simultaneously published in the USA and Canada"–T.p. verso.
Includes bibliographical references and index.
1. Twenty-first century–Forecasts–Congresses. 2. Social change–Forecasting–Congresses. 3. Power resources–Forecasting–Congresses. 4. Economic development–Forecasting–Congresses. 5. Ecology–Forecasting–Congresses. 6. Climatic changes–Forecasting–Congresses. 7. Land use–Forecasting–Congresses. 8. Sustainability–Forecasting–Congresses. 9. Earth–Congresses. I. Jäger, Jill. II. Cornell, Sarah. III. International Geosphere-Biosphere Program "Global Changes."
CB161.P57 2010
303.49–dc22 2010021498

ISBN: 978-0-415-59000-6 (hbk)
ISBN: 978-0-203-83812-9 (ebk)

Contents

Preface

Around 50 people from a wide range of backgrounds met in Lund, Sweden, in October 2008 to develop a vision for "the Planet in 2050" and discuss the pathways that could achieve that vision, as well as the challenges ahead. This book presents the results of that meeting. It describes a very different world in 2050 from the one we have today, as well as a number of major steps that would be required to change, amongst other things, the global economy, education, environmental governance and values and norms in society.

This book presents the outcome of a new kind of scholarly conversation, bridging academic disciplines and engaging with policy issues, practitioners and communities to create a collective vision for planet Earth in 2050. Continuing on a "business as usual" pathway for our collective human enterprise will lead to escalating society–environment problems, but agreeing on a desirable future presents enormous challenges. What constitutes a desirable future differs widely amongst individuals, societies and cultures, and multiple pathways are possible for getting from now to 2050 sustainably. Nevertheless, growing understanding of the fundamental processes that govern the functioning of the Earth system allows us to define the biophysical limits within which society can safely operate with more confidence. Within these limits, a rich array of desirable futures can be articulated, reflecting human creativity, diversity and determination.

Jill Jäger is an independent scholar living in Vienna, Austria. She has worked as a consultant on energy, environment and climate for numerous national and international organizations. Her main field of interest is the linkages between science and policy in the development of responses to global environmental issues.

Sarah Cornell is science manager and programme scientist for the QUEST (Quantifying and Understanding the Earth System) project based at the University of Bristol, UK. With a background in geochemical cycling (nitrogen) and atmospheric and marine chemistry, her interests have expanded to include sustainability science (optimizing the science–society interface for implementation of environmental policy) and the theory and methodology of cross-disciplinary integration.

1 The Lund meeting and the production of this book

Between 27 and 31 October 2008 a meeting was held in Lund and Malmö, Sweden, to develop a "vision for the Planet in 2050" and discuss how that vision could be achieved. The meeting was organized in two blocks, each of two days, with a "day off" in the middle of the week, during which the participants engaged in a variety of activities, including visits to sustainable development projects in the area and public events.

The meeting was organized by a small committee and supported by a wide range of institutions. Before the meeting, all of the invitees were asked to send "their vision" of the planet in 2050. These were made available on the meeting website (www.theplanet2050.org) and provided inspiration and background for the discussions that took place in Lund. Most of the meeting took place in working groups and was structured so there were opportunities for people to participate in the discussions of more than one group.

The Working Groups were:

- *Land-use change, agriculture and urban development* (chaired by Mary Scholes, Rapporteur Washington Odongo Ochola);
- *Energy and technology* (chaired by Martin Hoffert, Rapporteur Joshua Farley);
- *Development, global economy and culture* (chaired by Manuel Cesario, Rapporteur Sylvia Karlsson);
- *Environment, climate change, air and water pollution and health* (chaired by Sybil Seitzinger, Rapporteur Kathy Hibbard).

Each of these groups produced a written report, focusing on their vision for 2050 and the pathways to achieve that vision. After the meeting, as the editors began to put together the material from the meeting, it became clear, however, that it would be much better to restructure the four reports into the four following chapters of this book: the planet in 2010; the vision for 2050; the pathways to achieve that vision; and the challenges ahead. Thus, most of the material in this book is taken directly from the reports of the groups that met in Lund, but sorted into these chapters. In addition, excerpts from the individual vision statements have been included, to illustrate both the diversity of perspectives among the participants and, in some cases, very specific aspects of the visions.

1.1 The organizing committee

This group "met" regularly in telephone conferences organized so the massive difference in time zones could be bridged – thus while it was (very) early in the morning in Australia, it was (quite) late in the evening of the previous day in Europe and just after lunch in Colorado! There were also a few face-to-face meetings of sub-sets of the committee, but most of the discussion on potential participants, topics to be covered, logistics for the meeting, funding sources, etc. was carried out electronically and on the telephone.

The Committee was chaired by Guy Brasseur, and the other members were: Will Steffen, Torvald Jacobsson, Robert Costanza, Jill Jäger, Sarah Cornell, Rik Leemans and John Katzenberger.

1.2 The sponsors

The meeting in Lund was supported by a wide range of sponsors:

LMK industry manages financial holdings and provides strategic business development consulting to international growth-oriented high-tech companies. www.lmkindustri.com.

The Heinz Center is a non-profit, non-partisan think tank dedicated to improving the scientific and economic foundation for environmental policy. www.heinzctr.org.

National Center for Atmospheric Research provides the university science and teaching community with the tools, facilities and support required to perform innovative research. www.ncar.ucar.edu.

Quantifying and Understanding the Earth System (QUEST) is a programme created in 2003 by the UK Natural Environment Research Council with a focus on interdisciplinary research, closely targeted to help deliver a substantial improvement in our quantitative understanding of global environmental change. http://quest.bris.ac.uk.

Earth System Science Partnership (ESSP) is a partnership for the integrated study of the Earth system, the ways that it is changing and the implications for global and regional sustainability. www.essp.org.

The city of Lund is a meeting place for people and their ideas, knowledge and creativity. www.lund.se.

Lund University is an international centre for research and education that has approximately 38 000 students. www.lu.se.

The city of Malmö is the commercial centre of southern Sweden and an international city undergoing a transition from being an industrial city to a city of knowledge and sustainable development. www.malmo.se/english.

The International Institute for Industrial Environmental Economics at Lund University (IIIEE) seeks to facilitate the transformation of societies to sustainable development by utilizing the synergies of cross-disciplinary education and research that is based upon direct engagement with private and public sectors and government. www.iiiee.lu.se.

Region Skåne is a regional public body responsible for health, medical and dental services, and will also have responsibility for future issues. www.skane.se.

European Science Foundation is an association of 79 member organizations devoted to scientific research in 30 European countries and the core purpose is to promote high-quality science at a European level. www.esf.org.

The International Geosphere-Biosphere Programme (IGBP) is an international research programme that studies the phenomenon of global change. www.igbp.net.

Mossagården is one of the pioneering distributers and marketers of all organic fruits and vegetables in the south of Sweden. www.mossagarden.se.

1.3 Acknowledgements

This book is based on a rich discussion that took place in Lund and Malmö in October 2008. We thank all of the participants at the meeting for their contributions, written and spoken, to the development of a very ambitious vision of a peaceful, equitable and (becoming) sustainable planet in 2050. A list of all participants is provided below. We acknowledge with much gratitude the sponsors of the meeting, who ensured not only that participants could travel to Lund and stay in very pleasant surroundings, but also that they had opportunities to visit local points of interest that already contribute to sustainable development and engage in a dialogue with local actors interested in the future of the planet. All of our hosts in Sweden contributed to making the meeting a success, and without that success there would have been nothing upon which to base a book. In particular, we thank Torvald Jacobsson and Anna Gislén from Lund University for all of their efforts before and during the meeting. We also thank the other members of the Organizing Committee for their contributions before the meeting and for their support in producing this edited volume. Last, but by no means least, we would like to thank Elisabeth Knöppel of IIIEE in Lund and Barb Petruzzi of NCAR in Boulder, Colorado, for all of their support before, during and after the meeting: from the organization and running of the website, through the organization of the meeting itself, to the encouragement and enthusiasm as we worked on the book: without them it would have been much, much harder!

Figure 1.1 Participants at the Lund meeting, October 2008.

Note

At the time of the meeting, Lund was just starting its process for the city to become carbon-neutral, and all travel for the participants was compensated by tree planting.

1.4 Participants

Guy Brasseur *Associate Director at Earth and Sun Systems Laboratory, National Center for Atmospheric Research, Boulder, Colorado, USA*	**Jill Jäger** *Senior researcher at the Sustainable Europe Research Institute in Vienna, Austria*
Robert Costanza *Gund Institute for Ecological Economics, University of Vermont, USA*	**Rik Leemans,** *Environmental Systems Analysis Group at Wageningen Universiteit*
John Katzenberger *Co-founder and director of the Aspen Global Change Institute, Colorado, USA*	**Sarah Cornell** *QUEST, University of Bristol, UK*
Torvald Jacobsson *International Institute for Industrial Environmental Economics at Lund University, Sweden*	**Kerry Arabena** *Australian Institute of Aboriginal and Torres Strait Islander Studies, Australia*
Erik Assadourian *Worldwatch Institute, USA*	**Jayanta Bandyopadhyay** *Centre for Development and Environment Policy, India*
Aleh Cherp *Director for Research and Professor of Environmental Sciences and Policy at Central European University, Hungary*	**Manuel Cesario** *SAMAUMA/OASCA-UNIFRAN, Brazil*
Uchita de Zoysa *Centre for Environment and Development, Sri Lanka*	**Wenjie Dong** *Beijing Normal University, China*
Catherine Downy *QUEST, University of Bristol, UK*	**Malin Falkenmark** *SIWI, Stockholm Resilience Centre, Sweden*
Joshua Farley *University of Vermont, USA*	**Kathy Hibbard** *National Center for Atmospheric Research/AIMES, Boulder, Colorado, USA*

Martin Hoffert *New York University, USA*	**Thomas B. Johansson** *International Institute for Industrial Environmental Economics at Lund University, Sweden*
Sylvia Karlsson *Finland Futures Research Center, Turku School of Economics, Finland*	**Stefan Kaufman** *Environment Protection Authority, Victoria, Australia*
Ida Kubiszewski *University of Vermont, USA*	**Peter Marcotullio** *Hunter College, CUNY, New York, USA*
Michel Meybeck *CNRS/University of Paris, France*	**Jennifer Mohamed-Katerere** *Independent Expert, Member of IUCN Commission on Environmental Law and Commission on Environment, Economic and Social Policy, Jakarta, Indonesia*
Harold Mooney *Stanford University, California, USA*	**Washington Odongo Ochola** *Department of Agriculture, Education and Extension, Egerton University, Kenya*
Dennis Ojima *H. John Heinz III Center for Science, Economics and the Environment*	**Prof. Dr. Luiz Pingueli Rosa** *Federal University of Rio de Janeiro, Brazil*
Katherine Richardson *Faculty of Science, University of Copenhagen, Denmark*	**Tomas Ries** *Swedish Institute of International Affairs, Stockholm, Sweden*
Johan Rockström *Stockholm Environment Institute, Stockholm Resilience Centre, Sweden*	**Steve Running** *University of Montana, USA*
Roberto Sánchez-Rodríguez *Director and Professor of Environmental Studies at UC Mexus, University of California*	**Vernon Scarborough** *Department of Anthropology, Cincinnati, Ohio, USA*
Mary Scholes *University of the Witwatersrand, South Africa*	**Sybil Seitzinger** *Executive Director, IGBP Secretariat, Sweden*

Anthony Simon *Advisor to the President of the WBCSD, IIIEE Board*	**Stephen Simpson** *University of Bristol, UK*
Simron Jit Singh *Institute for Social Ecology, Faculty of Interdisciplinary Studies, Klagenfurt University, Vienna, Austria*	**Sofia Vaz** *Centre for Environmental and Sustainability Research, New University of Lisbon, Portugal*
Peter Victor *York University, Toronto, Ontario, Canada*	**Ajay Bhave** *International Institute for Industrial Environmental Economics, Lund University*
Anand Deshmukh *International Institute for Industrial Environmental Economics, Lund University*	**Jana Timm** *International Institute for Industrial Environmental Economics, Lund University*
Silvia Ceausu *International Institute for Industrial Environmental Economics, Lund University*	**Wayne J. Pan** *International Institute for Industrial Environmental Economics, Lund University*
Jessica Jewell *International Institute for Industrial Environmental Economics, Lund University*	

2 Developing a vision for the planet

2.1 Introduction

In 2004, the International Geosphere-Biosphere Programme (IGBP), together with the three other international global programmes that coordinate research on global change, published a synthesis of knowledge about global change – the profound transformation of Earth's environment as a result of the numbers and activities of people on the planet (Steffen *et al.* 2004). The synthesis showed the great acceleration of the transformations during the second half of the twentieth century. With the human population increasing from a little more than one billion up to six billion over the last 100 years, globalization of the economy and information flows and an increase in economic activity by a factor of ten between 1950 and 2000, human-induced environmental changes were becoming clear. These included land-use and land-cover changes, over-exploitation of fish stocks, changes to the composition of the atmosphere and loss of biodiversity. The synthesis also found that the magnitude and rates of human-driven changes to the global environment are, in many cases, unprecedented: "There is no previous analogue for the current operation of the Earth System" (Steffen *et al.* 2004, Preface).

The authors of the IGBP synthesis volume, together with a range of other (mainly) environmental scientists, were interviewed by a Swedish film crew between 2004 and 2006. This film crew worked for more than two years, visiting over 25 countries around the world to document the extent of global change and its underlying causes. The film, *The Planet*, was released in 2006 both as a cinema film and as a four-part documentary for television, and was shown in many countries. The visual style of the film, together with the use of music, gave an often breath-taking picture of the state of the planet.

While the book and the film both hinted at the need for action and some possible options for halting the acceleration of global change, they were mainly framed as a "wake-up call". By showing the extent of change but also the extreme complexity of the Earth system and the nature of the uncertainties, there was a risk that readers and viewers would respond by shrugging their shoulders and thinking that there is nothing that an individual or even societal groups can do to halt the transformation of the planet due to human activities. This small book is one attempt to respond to that fear. By creating a vision of the future in which human activities respect planetary limits, by showing possible pathways to achieve that vision in a way that also enhances human well-being for everyone, the participants of the workshop in Lund in October 2008 were attempting to show what is possible and to empower individuals and societal groups to take an active role in shaping the future of the planet.

2.2 Thinking about the future

Of course it is not new to think about the future. As Sondeijker (2009) points out in her book *Imagining Sustainability*, Thomas More's *Utopia* (1516), Aldous Huxley's *Brave New World* (1932) and Edward Bellamy's *Looking Backward: 2000–1987* (1887) are examples of previous attempts to think about the future. In recent years, in addition to a wide range of fictional portrayals of the future, approaches like foresight studies and scenario development and analysis have been increasingly used to explore possible developments in the future.

After the Second World War, scenario development and analysis were first used for military purposes (van der Heijden 1996). The value of scenarios was soon recognized by Herman Kahn and others who advanced the use of scenarios for other strategic planning applications (Kahn and Weiner 1967). The scenario approach was refined at Royal Dutch/Shell by Pierre Wack in the 1970s and 1980s, during which time Shell became a leader of the scenario approach to business planning.

Today, scenario development is used in a wide variety of contexts ranging from political decision-making (e.g. Kahane 1992; 1998) to business planning (e.g. Wack 1985; Schwartz 1996; Davis 1998) and from global environmental assessments (e.g. Gallopin *et al.* 1997; Cosgrove and Rijsberman 2000; Nakicenovic and Swart 2000; van Notten *et al.* 2003) to local community management (Peterson *et al.* 2003).

In 2002, the Global Scenario Group published a path-breaking set of scenarios that spurred the debate on the challenges of sustainability (Raskin *et al.* 2002). They considered three classes of scenario: conventional worlds; barbarization; and great transitions. Of greatest relevance to this book are great-transition scenarios, which envision profound historical transformations in the fundamental values and organizing principles of society. They assume that new values and development paradigms become the norm and emphasize the quality of life and material sufficiency, human solidarity and global equity, affinity with nature and environmental sustainability. These changes of values and development paradigms are held to be essential for a transition to sustainability. In most respects, the vision that is presented in the next chapter of this book belongs to the class of great-transition scenarios.

The scenarios of the Global Scenario Group were taken as the basis of the scenarios developed for the United Nations Environment Programme's third and fourth Global Environment Outlook reports (UNEP 2002; 2007), where the great-transitions class of scenario is represented by the "sustainability first" narrative (Box 2.1).

Box 2.1 The "sustainability first" scenario

A new environment and development paradigm emerges in response to the challenge of sustainability, supported by new, more equitable values and institutions. A more visionary state of affairs prevails, where radical shifts in the way people interact with one another and with the world around them stimulate and support sustainable policy measures and accountable corporate behaviour.

There is much fuller collaboration between governments, citizens and other stakeholder groups in decision-making on issues of close common concern. A consensus is reached on what needs to be done to satisfy basic needs and to realize personal goals without beggaring others or spoiling the outlook for posterity.

Source: Jäger *et al.* (2008)

With the increasing use of scenario development and analysis, there are also now numerous surveys of scenarios, their development processes and usage. Research projects are further developing the processes of scenario development and analysis and scenarios are being used in large-scale assessment processes such as those of the Intergovernmental Panel on Climate Change (IPCC) and the Millennium Ecosystem Assessment.

2.3 The value of scenarios

It is not possible to predict the future. Our knowledge about all of the interacting processes that lead to continual change in the Earth system is incomplete. Predicting human behaviour, choices and changes of human values is clearly not possible either. Even if precise information on the Earth system were available, it is known that complex systems exhibit turbulent behaviour, extreme sensitivity to initial conditions and branching behaviours at critical thresholds, which all make prediction impossible. Given that prediction is not possible, scenario development and analysis offers a means of exploring a variety of long-range alternatives.

Useful scenarios are carefully created stories about the future. They can include both qualitative (i.e. a narrative) and quantitative aspects. While some scenarios have been (and are being) developed through very participatory processes, others are created in more "expert-driven" processes. Scenarios explore *what* could happen *if* particular developments come to pass (forward-looking) and they can also be developed to illustrate the paths that could lead to particular outcomes (back-casting).

The insights provided by scenario development and analysis processes can be used to support decision-making. An interesting and inspiring example of this is the "Mont Fleur" scenario exercise carried out in South Africa in 1991–92. The purpose of the exercise was to stimulate debate about how to shape policy in South Africa over the next ten years. The exercise was carried out by a diverse group of 22 prominent South Africans (e.g. politicians, activists, academics and business people) to develop and disseminate a set of stories about what might happen in their country over the next decade. At a time of deep conflict and great uncertainty in the country, the exercise was extremely important in stimulating the varied participants to think creatively about the future. First published in 1992 in the South African newspapers, *The Weekly Mail* and the *Guardian Weekly*, the scenarios were widely publicized.[1]

The participants agreed on four scenarios that they believed to be plausible and relevant:

> **Ostrich**: a negotiated settlement to the crisis in South Africa is not achieved, and the country's government continues to be non-representative.
> **Lame duck**: a settlement is achieved but the transition to a new system is slow and indecisive.
> **Icarus**: transition is rapid but the new government unwisely pursues unsustainable, populist economic policies.
> **Flight of the flamingos**: the government's policies are sustainable and the country takes a path of inclusive growth and democracy.

The scenarios were discussed widely. Through this process, it became clear that **flight of the flamingos** was a feasible and broadly desirable outcome, although some of the decisions it implied were not in line with those that might have been proposed by some of the parties at the start of the exercise. Thus, the informal process of producing the Mont Fleur scenarios produced substantive messages, informal networks and changes in thinking about the challenges that the country faced.

As illustrated by the example of the Mont Fleur scenarios, scenarios support informed action by providing insights into the scope of the possible. In the process of helping to clarify possible future developments and their effects, scenarios are often a source of inspiration for creative ideas.

Scenarios can be and are used for multiple purposes, including to: stimulate discussion and creative thinking; understand the world better and make better decisions; encourage social learning between and within different groups of actors; test strategies with respect to their robustness; and raise awareness (for example about the role of human activities in shaping the future and/or the linkages between issues).

2.4 Visions

As described above, **scenarios** are consistent descriptions of a possible future situation as determined by those factors that are **both most important** and **most uncertain**. The processes used to develop scenarios (see, for example, Jäger *et al.* 2008) include consideration of the driving forces of change and a selection of the most important and the most uncertain factors. A scenario is a plausible story about how the world might evolve in the future.

A **vision** (see, for example, Moriarty *et al.* 2005) is a concise description of a desired future state. Visions provide a picture of *how* we would like the world to be at some future time. Visions have been and are developed in a wide range of projects and processes. A step of "envisioning" is included, for instance, in integrated sustainability assessment (Weaver and Rotmans 2006).

The big difference, therefore, between a scenario and a vision is the word "desired". In a vision, the developers are considering how they would like the world to be. A powerful call for the need for visions was made by Donella Meadows (1996):

> Vision is the most vital step in the policy process. If we don't know where we want to go, it makes little difference that we make great progress. Yet vision is not only missing almost entirely from policy discussions; it is missing from our whole culture. We talk about our fears, frustrations, and doubts endlessly, but we talk only rarely and with embarrassment about our dreams. Environmentalists have been especially ineffective in creating any shared vision of the world they

> are working toward – a sustainable world in which people live within nature in a way that meets human needs while not degrading natural systems. Hardly anyone can imagine that world, especially not as a world they'd actively like to live in. The process of building a responsible vision of a sustainable world is not a rational one. It comes from values, not logic. Envisioning is a skill that can be developed, like any other human skill.

A vision, as Meadows emphasized, is not about the world that people think they could achieve, nor is it about the world they would be willing to settle for, it is about the world they truly want. Like a scenario, a vision has the power to open the mind to possibilities that would not be seen in a mood of cynicism. A vision makes it clear that there is a wider range of choices that can be made in the future and it can unleash the kind of creativity needed to achieve a desired future state. It can also show where there are currently seeds of good practice, which, if nourished, can grow and be part of achieving the desired state.

Of course, as Meadows and others have underlined, simply having a vision is not enough. It has to be followed by a process of discussing the ways to achieve the vision and a process of implementation, continuous evaluation, adjustment of the goal on the base of experience and so on. This is the process used, for example, in transition management (Rotmans *et al.* 2000).

Various efforts have been made nationally and worldwide to develop collective visions of sustainable futures through participatory engagement processes. Some have taken a long-term and large-scale perspective, engaging community leaders (e.g. the Australia 2020 Summit[2]; Envisioning a Sustainable and Desirable America[3]). Others have been more locally oriented, community-engaged and focused on shorter-term decision-making, such as the Neighbourhood Initiative Foundation's Planning for Real[4] and the United Kingdom's Quality of Life Capital.[5] The World Business Council for Sustainable Development has also developed a vision for 2050[6] that lays out a pathway leading to a global population of some nine billion people living well, within the resource limits of the planet. This vision was the result of an 18-month combined effort with CEOs and experts, and dialogues with over 200 companies and external stakeholders in some 20 countries. Some visions focus on a particular sector, such as energy. *A Renewable World*[7] was developed in conjunction with dozens of world experts, and features solutions on renewable energy, biosequestration, energy sufficiency, energy for developing countries, green-collar jobs, cities, transport, agriculture and food, regional economies and civil participation and democracy.

A recent vision for the European Union (*Getting into the Right Lane for 2050* (PBL 2009) considers, in particular, three key challenges: producing food to feed a global population of nine billion while minimizing biodiversity loss; mitigating climate change while enhancing energy security; and developing a low-carbon transport system in the EU. On the basis of the vision, it was concluded that strategic actions for the EU in the coming 5–10 years are decisive for the achievement of the vision. A further activity for Europe that is relevant to readers of this book is the *Roadmap 2050*, which examines scenarios for a low-carbon Europe.[8] One main conclusion of the *Roadmap 2050* is that decarbonization by 2050 demands strong action before 2015.

2.5 Conclusion

This book presents a vision for the planet in 2050. That is, the participants in the workshop in Lund created a collective picture of a desired state in the future. This desired state stands in stark contrast to the world that we currently live in with respect to the environment, economy and society. It should not be dismissed as "utopian", "crazy" or "unrealistic". As Donella Meadows points out, we need a vision in order to develop a plan for a more sustainable future, to inspire us to think more creatively about the options that are available to us and to show us which current practices need to be nurtured as well as those that are problematic.

3 The starting point

Planet 2010

3.1 Introduction

Before developing a vision for 2050, it is important to take stock of the current situation. We first describe the state of the environment, focusing on the unprecedented rates of change of the past 50 years or more, the changes of ecosystems upon whose services human life on this planet depends, and changes of climate. We discuss the fact that many of the changes are interlinked. Air pollution and problems with water quality and quantity are then discussed, since these issues are persistent problems faced in many parts of the world in 2010.

> Massive inequalities in access to energy services, billions of people without access to electricity or clean cooking fuels;
> Indoor and outdoor air pollution from dirty fuels and associated 5% of the global burden of disease;
> Crumbling and unreliable energy infrastructure jeopardizing normal functioning of societies;
> Gunboat diplomacy, energy blackmail, geopolitical tensions, and "oil wars";
> "Resource curse" crushing political and economic hopes of energy-exporting nations.
>
> Aleh Cherp

Following this review of the environment in 2010, we retrace our steps and review how we got to this situation and how the relationship between nature and society has changed. This leads to some discussion of society in 2010 – governance, the state of the global economy, the role of beliefs and values and the quality of life. The chapter ends by synthesizing the state of the environment and society in 2010 in order to set the stage for the vision for 2050 and the pathways to reach that vision.

3.2 The state of the environment

We are in an era of unprecedented rates of human-induced environmental changes with unknown rates of Earth system response. Figure 3.1 is an illustration of rates of change in the climate system, but there are similar images of accelerating changes for many parts of the Earth system (see, for example, Steffen *et al.* 2004). The symptoms of our malfunctioning biosphere are clear: diminishing resources, compromised air and water quality, alarming rates of biological extinction and declining biodiversity, and a climate system that is changing in response to hitherto unprecedented rates of change to the atmosphere. To a large extent, recent changes are well understood and are largely attributed to the human enterprise, including industrialization, massive changes in transportation infrastructures, deforestation, large-scale agricultural practices and their associated fertilization and irrigation practices.

3.2.1 Changes of ecosystems

The Earth is inhabited by millions of different kinds of creatures, each uniquely adapted to their local environment in order to acquire the energy, as well as the other vital resources, necessary to sustain their growth and reproductive capacity. Through reproduction, the genetic information gained during the course of evolutionary processes is passed on to future generations. The struggle to obtain energy, water and nutrients is true of all life. It is the competition to capture these resources, wherever they might reside, that leads to the great diversity of life and the complex food chains that characterize it.

The assembly of species and their interactions with each other, and with the physical environment, constitute an ecosystem. These interactions result from the mining of resources by organisms from the environment. They do this by utilizing the energy captured directly from the sun or by consuming other organisms. Primary producers in terrestrial systems pull water and nutrients from the soil. The water moves through plants to the atmosphere, unavoidably being lost from the plant due to the requirement for capture of CO_2 from the atmosphere. These ecosystem processes, the capture, utilization and loss of energy, carbon, water and nutrients, form the basis for the human life-support system, and the flows of water and nutrients represent the operation of biogeochemical cycles. In an analogy, ecosystem processes can be considered a factory. As in any factory, there are many components, or, for our analogy, species, but these pieces are of different sizes and abundance and they perform different roles, again as in an ecosystem. The difference in an ecosystem is that the parts are

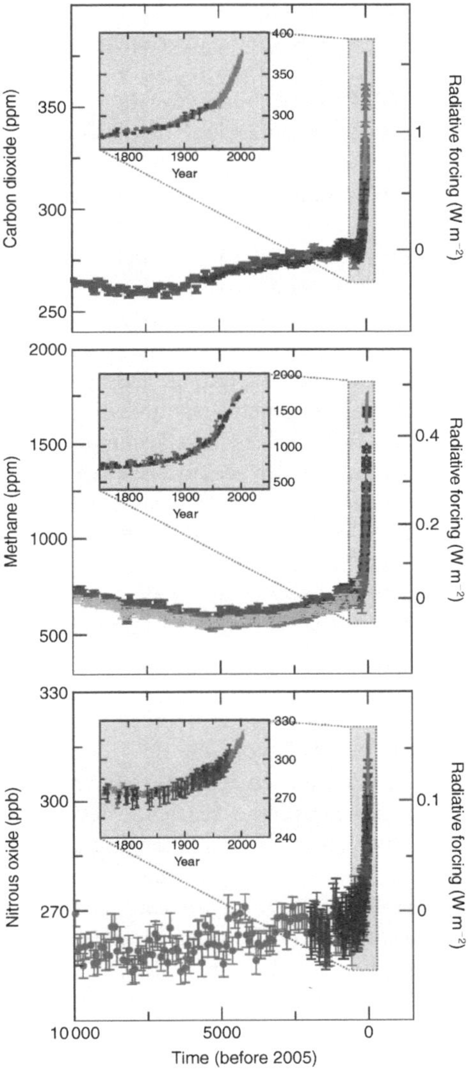

Figure 3.1 Atmospheric concentrations of carbon dioxide, methane and nitrous oxide over the last 10000 years (large panels) and since 1750 (inset panels) (source: IPCC 2007).

Note

Measurements are shown from ice cores and atmospheric samples. The corresponding radiative forcings are shown on the right-hand axes of the large panels. Ppm and ppb stand for parts per million and parts per billion, respectively.

self-replicating, and further, new species can enter into the system and alter the whole dynamics of the system by how they gather energy, water and nutrients.

The functioning of ecosystems and the services they provide are essential to society and its well-being. Ecosystem services include provisioning of food and fibre, or provisioning services; regulating services, such as climate regulation, flood control and pollination; and cultural services such as recreation, aesthetic beauty and spiritual inspiration.

In 2005, the Millennium Ecosystem Assessment, which was five years in the making, concluded that during a very brief period in history – the past 50 years – we have seen a dramatic reduction in the capacity of the Earth's biotic systems to deliver ecosystem services. We have degraded some 60 per cent of nature's ecosystem services, including water purification, disease and pest regulation, pollination, genetic resources, fisheries and aesthetic values, to name but a few. These losses have accumulated as we have dismantled the base of the chain – we are seeing losses of diversity in all dimensions, from genetic diversity, species richness, population sizes and ecosystems. With these losses, the delivery of the services of the "factory" goes with them. Moreover, the Millennium Ecosystem Assessment concluded that these losses are accelerating and that the next 25 years will most likely see losses as great as we have seen in the past 50 years.

3.2.2 Changes of climate

As shown by the Fourth Assessment Report (AR4) of the IPCC, observations of increased global average air and ocean temperatures, widespread melting of snow and ice, and rising global average sea level have confirmed that warming of the climate system is unequivocal (Figure 3.2). In addition, the AR4 confirmed that most of the observed increases in global average temperatures from the mid-twentieth century are very likely due to the observed increase in anthropogenic, or human-caused, greenhouse gas (GHG) concentrations in the atmosphere (Bernstein *et al.* 2007).

In recent years, there has been an *increasing* growth rate of CO_2 emissions, propelled by society's apparently insatiable demand for energy. Figure 3.2 from Raupach *et al.* (2007) shows the actual emissions (black line) as well as emissions scenarios for the twenty-first century (lines in various shades of grey) from the IPCC. This growth in emissions has led to observed changes in the global average temperature, sea level and snow cover in the northern hemisphere (Figure 3.3).

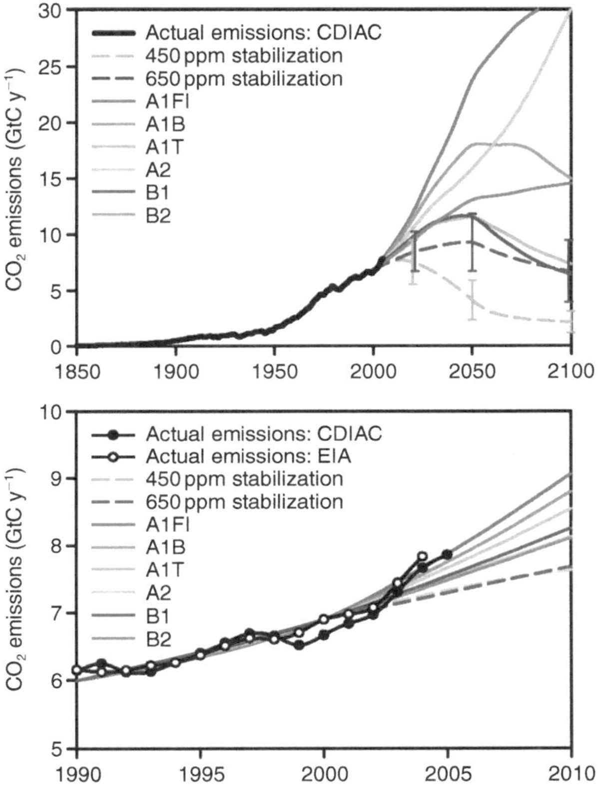

Figure 3.2 Observed global CO_2 emissions compared with emissions scenarios and stabilization trajectories leading to greenhouse gas stabilization at two different levels (dashed lines; present day atmospheric concentrations are 392 ppm) (source: Raupach *et al.* 2007).

Note
Six example IPCC scenarios are shown, giving possible future emissions associated with different storylines for socio-economic development. The *A* scenario family emphasizes globalized development, while the *B* scenarios emphasize regionalized development. Similarly, the scenarios numbered *1* prioritize economic growth, while those numbered *2* place higher emphasis on environmental stewardship. Three variants of the A1 (globalized, economically oriented) scenario lead to different emissions trajectories: A1FI (intensive dependence on fossil fuels), A1T (alternative technologies largely replace fossil fuels) and A1B (balanced energy supply between fossil fuels and alternatives).

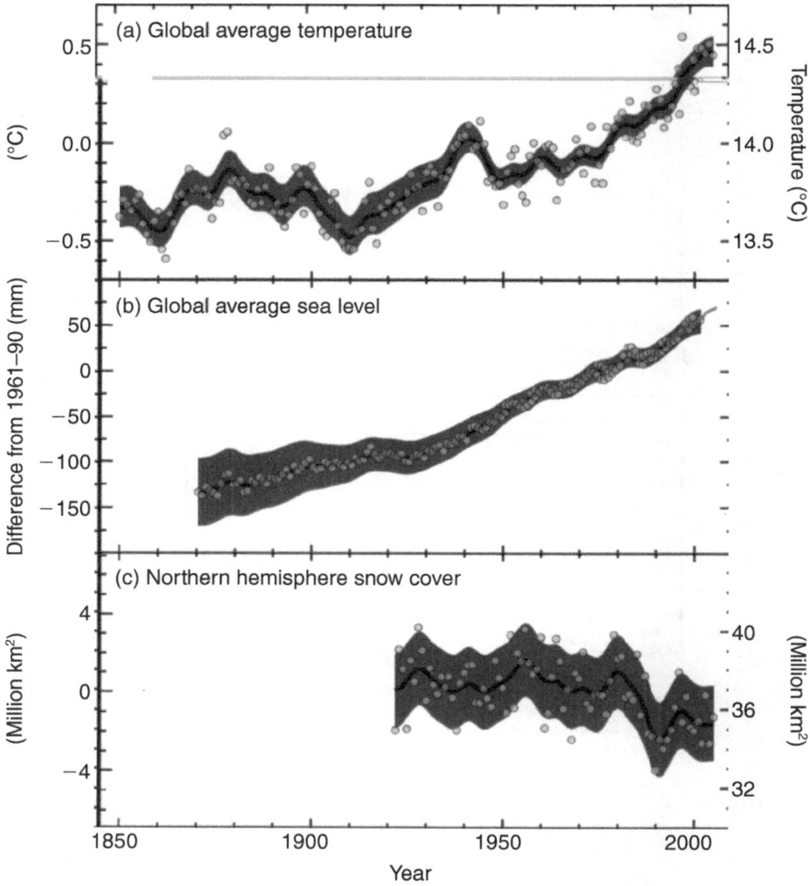

Figure 3.3 Observed changes in (a) global average surface temperature; (b) global average sea level from tide gauge (black) and satellite (light grey) data; and (c) northern hemisphere snow cover for March–April. (source: Bernstein *et al.* 2007).

Note

All differences are relative to corresponding averages for the period 1961–90. Smoothed curves represent decadal averaged values while circles show yearly values. The shaded areas are the uncertainty intervals estimated from a comprehensive analysis of known uncertainties (a and b) and from the time series (c).

3.2.3 Multiple and interacting changes

By the early years of the twenty-first century, there was widespread recognition that human beings, as individuals and as societies, interact with the environment in complex, dynamic ways. Rather than being controlled by simple cause-and-effect relationships, society's pathway is often the result of multiple interacting factors driven by environmental and socio-political changes that may manifest themselves at various scales. Research studies have highlighted, for example, the complex interactions of agriculture, land use, water resources and biodiversity with climate change (e.g. Backlund *et al.* 2008). Lessons from the past can often provide insight into current diagnostics of such social and environmental interactions (e.g. Hegmon *et al.* 2008, 1998; van der Leeuw 2005).

One area in which the multiple and interacting pressures are starkly manifested is in cities (see Box 3.1). Urbanization and cities are sources of economic growth (Jacobs 1969; World Bank 1999). Urbanization is a natural concomitant of modernization and industrialization (Bradshaw and Fraser 1989). Satterthwaite (2007) has pointed out that around 97 per cent of the world's GDP is generated by industry and services, and around 65 per cent of the world's economically active population works in these sectors. Both industry and services along with these populations are located largely within urban areas.

Box 3.1 Urbanization and multiple and interacting pressures

Urbanization is intimately intertwined with a number of other social and biophysical processes and itself is both an impact and a driver of change. In terms of direct influence, it is difficult, if not impossible, to tease out the individual influence of urbanization from these other processes.

In 2010 more than half of the world's population lives in cities. The proportion of the world's population that lives in megacities (a metropolitan area with a total population in excess of ten million people) is merely 4.5 per cent, or 9.3 per cent of the global urban population. From 1950 to 2005, the number of megacities rose from 2 to 20. However, more than 50 per cent of the urban population lives in cities of 500000 people or less.

Cities have long been associated with some of the worst outfalls of development, both social and environmental. Brown and Jacobson (1987) state that those living in cities create a disproportionate share of waste, land and water scarcity and engage in inefficient energy usage. There are continual fears that large cities are becoming overrun, chaotic and unmanageable places (Linden 1996).

Cities are increasingly sources of diverse technologies, activities and people (Amin and Graham 1999). This diversity, however, should not be confused and equated with loss of values or social dysfunction. While within any city around the world there are families with non-traditional structures or values and there are homeless individuals, there are also true neighbourhoods and communities, and those that cling to more traditional ways of life (Smart and Smart 2003). Cities are complex phenomena and can be sources of hope and inspiration.

While cities are indeed growing rapidly, the growth of megacities is slowing. Some of the most rapid growth of urban centres is being experienced by cities in North America (Satterthwaite 2007). Interestingly, when compared to the developed world, many emerging economies and poor countries are not only growing more efficiently, say in terms of energy use per capita (Marcotullio and Schulz 2007), they are also well run.

The impact of urban development on various environmental media varies dramatically with the design and governance of the city (McGranahan *et al.* 2005). Suburban areas have much lower levels of hard surfaces. While some biodiversity is highly affected (the built environment particularly impacts large, animal "urban avoiders"), plant diversity (while different from indigenous associations) has been documented to increase within parts of some cities (McKinney 2008). This is not to suggest that urban built environments are good for the local issues of biodiversity, soils or hydrology, but rather that they do not necessarily need to be as negative as some are.

3.2.4 Air pollution

Air pollution has become a problem of global significance with important health consequences. The primary determinants of air pollution are the scale and composition of economic activity, population, the energy mix, the strength of local pollution regulation, and geographic and atmospheric conditions that affect pollutant dispersion in the atmosphere. Even though measures have been taken to limit emissions of primary pollutants in industrialized regions of Europe and North America, the number of days exceeding the air quality standards in these regions remains excessive. At the same time, economic growth in developing countries has created huge amounts of pollution (along with wealth), both in urbanized centres and rural areas. Air quality in cities in China, Africa and South America has worsened dramatically during the last decades, with considerable impacts on health and, according to the World Health Organization, the premature deaths of nearly one million people each year. Reduction in crop productivity and impacts on natural ecosystems are also considerable. The reliance of several countries on coal (an extremely dirty fuel source), the extensive shift towards diesel automobiles and insufficient environmental regulations are also to blame.

3.2.5 Water quantity and quality

> Water is the most fundamental component for life. It is more primary than food. As a proxy for all our most primary of resources, how we treat it reveals our attitudes towards the planet and all those that occupy its habitats. We need to treat it better, and by doing so we will treat ourselves better.
>
> Vernon Scarborough

Rivers continuously convey fluxes of water, energy, sediments, carbon, nutrients (nitrogen, phosphorus, silica) from land to the coastal zone and to inland seas. The balance of these fluxes is fundamental to coastal sustainability, at medium time scales (10–100 years) and to the sustainability of regional seas (100–1000 years) and the global ocean (>1000 years). Riverine fluxes may either be greatly increased by human activities (e.g. some nutrients have globally increased by a factor of 3; regionally by a factor of 10; and locally by a factor of 50 or more), or greatly reduced by damming and water-consumptive use and diversion. In the latter case, rivers may not flow at all during the dry season, or their fluxes can be completely (>90–99 per cent) stopped, as has happened with the Colorado, the Nile, the Indus and many other rivers on all continents. The resulting nutrient depletion of coastal waters can generate a decline in coastal fisheries; the sediment depletion generates delta subsidence – hence seawater intrusion and increased risk of flooding caused by sea-level rise – and coastline erosion in sediment-starved areas. Rivers where the water discharge has been significantly decreased are much more sensitive to water quality issues. They also experience the greatest loss of aquatic biodiversity.

Rivers are an illustrative example of the interlinked socio-ecological system of which we are part. Because aquatic systems provide so many essential environmental benefits, human settlements were often established along rivers, and attempts were made to manage the rivers in order to exploit those benefits. The biodiversity and functioning of many of the world's aquatic ecosystems have become deeply linked to human activities, evolving with human development according to the multiple pressures that gradually developed in a river basin, on a river reach or at a river station. Sometimes external drivers may generate new pressures on a river system, such as long-range atmospheric pollution (e.g. acid rain, nitrogen fallout) and, more recently, global climate change. From an Earth system perspective, the key human-induced issues are:

- river habitat degradation (e.g. canalization, navigation locks, etc.);
- toxic pollutant inputs;
- nutrient inputs (nitrogen, phosphorus) and inputs of biodegradable organic matter resulting in eutrophication (excess growth of algae, which depletes oxygen from the water when the algae die and decay);
- loss of longitudinal continuity (damming);
- over-extraction of water (for consumption, irrigation and so on) – river discharges can reach seasonal or complete dryness, causing total loss of aquatic biota and partial loss of riparian biota.

In addition to these issues that are linked to land use, water use and urbanization within the river basin, climate change adds another set of pressures on the aquatic biota (e.g. rainfall reduction, temperature increases in rivers, melting of glaciers in high-altitude headwaters). The timing of these issues is specific to each river basin. Some rivers have already been impacted for hundreds of years or more, for others the impacts have just started.

> Rivers globally have evolved much faster in the past 50 years than in the last 5000 years, mostly due to direct human impacts. In the next decades, they will be exposed to: (1) evolving water demand due to population and land-use growth; (2) regulation policies of both water and land use; (3) technological developments and innovations; and (4) global climate change. The combined effects of these drivers will be very different on each river basin: at the global scale they can be described as a giant puzzle of a few hundred pieces that evolve separately from each other.
>
> Michel Meybeck

Water quality

In the history of human development, river water quality issues have gradually occurred through stages from very limited impacts for hunter-gatherer societies to maximum impact degradation found in impounded rivers, combining developed irrigation, megacities, industries and/or mining activities with limited waste-water collection and located in countries with limited or unenforced environmental regulations, absence of environmental education and limited citizen awareness.

Stage 1: population-driven situations. The water use and the pollution loads (e.g. per capita uses and loads) are nearly constant: the population is the main driver of the pollution, which depends on the capacity of the receiving water body, dilution and self-purification of organic pollution.

Stage 2: accelerated degradation. At some point the pressures progress faster than the population increases. This occurs either through the increasing per capita consumption (food, energy, goods) with a high leakage rate to the aquatic environment (nutrients, metals, pesticides, industrial and mining wastes) or through the collection of waste-waters as opposed to individual sanitation without related treatment. The river reaches downstream of point sources are the most severely degraded (e.g. anoxia, very high bacterial counts, very high concentration of toxins). This degradation is greatly amplified by decreased river runoff.

Stage 3: environmental recovery and river basin governance. The population growth rate has decreased or is stabilized. The society has developed infrastructures for proper waste-water treatment; industrial and mining wastes are recycled; use of toxic substances (e.g. mercury, cadmium, PCBs, DDT, atrazine pesticide) is very much reduced or banned; environmental education and citizen awareness are effective and raise the acceptance of strict environmental regulations and the cost of infrastructures; the river basin (or regional sea) governance harmonizes regulations and sets common targets and compensation mechanisms.

Stage 4: fully managed river quality. Water quality goals are a societal priority. The water system can be directly engineered to maintain the quality for users and aquatic biota. All dams are equipped with fish ladders, some of which can be decommissioned; drought events associated with poor water quality are compensated by upstream reservoir water release. During rainstorm events in megacities, oxygen is injected directly into the river downstream of combined sewer overflow. Surface runoff waters are stored to prevent inundations and water quality degradation.

These stages greatly depend on:

- the state and rate of human development within the river basin (which may be heterogeneous);
- the political priority given to the water resources and to the aquatic environment;
- the awareness and consensus among all common governance bodies (e.g. river basin agencies, regional seas treaties).

The least developed countries are generally at Stage 1. The fast-developing countries went from Stage 1 to Stage 2 in the last few decades. Some of the Eastern European countries mostly developed under Stage 2, until the early 1990s. Most countries in North America and Western Europe went from Stage 2 to Stage 3 during the post-Second World War period. In these regions, local river basins are now in Stage 4.

Proper management of river basins and water use can generate river recovery and the restoration of its aquatic life. Yet it is generally not possible to go back to a pristine state and to the original biodiversity, for instance when endemic species are lost or when there are species introductions and invasions.

Water quantity

The availability of water per person is diminishing, while expectations of a better quality of life are increasing in developing countries (SIWI 2007). The quality of much of today's water is poor. Most of today's undernourished people live in dry climate regions in South Asia and sub-Saharan Africa. There are 2.6 billion people still lacking access to basic sanitation.

In the past 20 years or so, there have been changes in the climate conditions over large regions. The resulting green water shortage (see Box 3.2) has caused reduced food production in affected regions and ecosystem stress. Also, aquatic ecosystems are now being impacted by declining late-summer flow-rates in semi-arid climates.

Box 3.2 Water: bloodstream of the biosphere

Life in the biosphere is intimately linked to the presence of water and to temperatures that allow this water to exist as ice, liquid water and vapour. The energy absorbed and released in the transformations from one of these three states to another plays major roles in the redistribution of solar energy around the surface of the Earth. Water is also unique as a solvent, dissolving matter and releasing it as it moves around the planet. This is the way plants obtain nutrients from the soil, by water sucked in through the roots and up through the stem – a process driven by evaporation from the leaves when stomata open to take in one of the raw materials for photosynthesis, carbon dioxide from the air. But water itself also has the function of raw material, providing building blocks for the process which starts by the splitting of the water molecule.

Water solubility also plays fundamental roles in the human body, most of which – some 70 per cent of the body weight – is water. It carries nutrients to and waste away from the cells. The body loses water by sweating and excretion of urine, but has to retain its normal water content by daily water intake – fundamental for survival since the body functions cannot endure more than a limited water loss. Hence the importance of safe drinking water; safe in the sense of freedom from disease-generating matter.

Globally, the water keeps continuously cycling between ocean, atmosphere and continents. A landscape gets its water by precipitation, partitioning it between "green" water in the upper soil, picked up by the plants and returned to the atmosphere, and liquid "blue" water, moving through rivers and aquifers by gravity towards the sea or, in arid regions, to a closed lake.

It is in these landscape-scale surroundings, with their circulating water, that ecosystems have developed and that humans are living their lives. Terrestrial ecosystems with their trees, grasses and flowers feed on "green" water from the soil; aquatic ecosystems with their algae and fish dwell in liquid "blue" water surroundings, forming their habitats.

Human societies have, since ancient times, been using the liquid "blue" water for many parallel purposes: water supply of households and communities, industrial production purposes, hydropower production and cooling, irrigation in arid regions, transpiration, etc. But they have also used it for carrying away the waste-water after use, resulting in an increasing pollution of water courses as populations have grown. Where climates have changed and severe droughts set in, societies have vanished.

As population grows and socio-economic development proceeds, water demands have kept growing. Direct human needs are small on a relative scale, and are basically a problem of finding an acceptable raw water source. As water gets scarcer and pollution regulations sharper, industry increasingly re-uses its water. Competition grows between irrigation – the dominant water use – and leaving water in the river for downstream aquatic ecosystems. An increasing number of water-scarce countries therefore have to apply re-allocation of available water.

3.3 Retracing our steps: how did we get to 2010?

From the earliest times of human settlement, *homo sapiens* have a long history of environmental degradation and over-exploitation of resources (e.g. Hegmon *et al.* 2008). Loss of environmental resources prompted various responses, from the ingenious invention of irrigation, to migration of communities, to the disintegration and collapse of whole societies. This legacy is not unique to any single settlement or culture. To acknowledge this feature of humans and their interactions with the environment from the outset is necessary for any critical evaluation of how we must proceed towards any future. The critical difference between environmental circumstances now and 10 000 or even 200 years ago is the rapid growth in human population and material consumption, and the acknowledgement that we have reached a limit where the Earth system now has a diminished capacity for absorbing human-induced changes and waste products.

Our deepest origins as cultural beings are not necessarily an obvious outgrowth of our bipedal gait or even our highly visible and hallmark evolutionary signature of stone tool-making – i.e. technology. These attributes probably appear on our timeline between two million and three million years ago. An important contributor to becoming a successful species was using our big brains to harness fire, perhaps 500 000 years ago. The daily activity of cooking did indeed promote our further separation from our immediate ancestors, accommodating our physical alertness and increasing attraction to meat sources and the proteins that fuelled our growing and convoluted brain structures. But fire more fundamentally lengthened our days and provided the opportunities for occupying the Earth's varied temperate setting, by artificially controlling heat and creating immediate micro-environments for social cooperation and coordination of resources.

Through a series of complex and poorly understood pathways, a version of our present subspecies evolved and likely migrated "out of Africa" by around 100 000–200 000 years ago. The principal facilitator of our cultural definition – language – was well established no later than 40 000 years ago, a biological breakthrough that significantly enhanced our potential for symbolism, learned behaviours and social action. By 10 000 years ago, the resource abundance in certain pockets of the Old World, best noted in the Levantine hills of the Near East, began to accommodate sizable populations of less mobile hunters, gatherers and foragers that began to experiment with wild grains and related foods, storage and incipient domestication. The role of agriculture was the pivotal circumstance that changed our economy and our ecology from one of low-impact collecting to high-impact food production. And our planet has never been the same, nor has our resilience as a species.

With the advent of agriculture, human societies have had major impacts on biophysical relationships for at least the last 10 000 years. With the appearance of cities at least 4500 years ago, the tremendous impact of our engineered landscape began to take shape, and our precipitous societal investments in structural or institutional development accelerated. With the advent of the industrial revolution, our ability to exploit the environment and one another took yet another turn. Accumulation of wealth was made possible with the introduction of fossil fuels and its supportive technology, but this wealth only benefited some countries. Resource consumption as appropriated from the biophysical environment intensified and accelerated exponentially, yet only a small portion of the world's population participated in this consumption growth. Poverty and under-consumption remained ingrained within most countries of the world, preventing people from having access to food, shelter, health, education, security and influence

As a species, we have been vulnerable throughout history at times of change in the surrounding environment, but our cultural flexibility and creativity have allowed the kinds of resilience necessary to produce new socio-ecological adaptations – for example, through plant and animal domestication. However, business-as-usual since the advent of agriculture was not a sustainable trajectory – exploitation and its polluting consequences heightened the need for structural change in the human trajectory. Humanity had developed different types of social structures, formal and informal institutions within the economic, political and social domains, but the type of culture we shaped has contributed to and supported the unsustainable trajectory.

3.4 The changing relationship between society and nature

One way of looking at the environmental situation in 2010 is as an imbalance in a coupled social and ecological system. Looking at the whole world as a complex adaptive system, where the physical environment, nature and human society are fundamentally interlinked, is proving a very useful conceptualization (e.g. Olsson *et al.* 2006). The complexity of the system arises from the intricate interplay among living organisms, including the human species, and their interactions with their physical "home", the planet Earth. The complexity of the system means that changes in one aspect often result in effects felt elsewhere. Human consciousness extends the normal conception of adaptive systems, where the current dynamics are simply shaped by the historic development of the system, to one where there may really be a capacity to learn from experience and select deliberately among potential future pathways.

It has become clear that the current pathways of consumption cannot be continued into the future: the natural system constrains the socio-economic subsystem (Daly 2005). This constraint has been disregarded so far, partly because it is far from clear where the limits could and should lie. Limits get recognized only when one is inadvertently exceeded. In the past, such a lack of recognition of the limits has led to the collapse of some human communities, including whole civilizations at a significant regional scale (Diamond 2005; Ponting 2007). The consequences of a change to this interconnected system – perhaps the loss of a single species, or the reduction of a key natural process below a threshold – could have knock-on impacts for our overall life-support system. There is now a need for a new focus on identifying what might be the boundaries and safety margins in human–nature relations. In the absence of this, human civilization could be putting itself at risk (where the precise cause and the magnitude of that risk may both be unknown).

There is no shortage of evidence that nature influences the working of human society, and that society shapes nature to its advantage. In other words, changes permeate the interconnected system. Human society has evidently exploited this throughout its existence – agriculture and technological change are adaptations that have extended the reach of human society across the world (Hobbs *et al.* 2006).

There are still comparatively few conceptual tools to describe and explore this socio-ecological system as an integrated whole. In most cultures worldwide, academic divisions have fallen between the study of human society and the study of the physical and natural world. As a result of this divide, there is now a profound epistemological and ontological divergence – social scientists and natural scientists quite literally see the world differently and speak different languages. The incompleteness of expert knowledge about the interconnected nature–society system is one constraint on effective response (Toffler 1984). The development of discipline-based knowledge has happened over the last 300 years or so, but there is arguably a need to accelerate the process of their reintegration for the generation of multidimensional, holistic knowledge.

There are some examples of this conceptual integration. In the area of environmental policy, the Drivers–Pressures–State–Impact–Response (DPSIR) framework adopted by the European Environment Agency captures the iterative dynamic of human and environmental change. Societal driving forces change the state of the environment, eventually impacting on the human societies that depend on a well-functioning natural environment to the point that they seek responses, which can involve direct remediation of the problem, mitigation of its effects, or more fundamental adaptive changes to society's priorities. With regard to environmental management, there is also well-tested operational experience – particularly in coastal zones and river basins, where environmental problems have become profoundly evident and the recognition of the diversity of stakeholders (including flora and fauna – non-human stakeholders!) compel more integrative dialogues.

This rebalancing of the interaction of our own species with the rest of nature will depend on a reversal of a very widespread and long-standing detachment of humans from nature. Some cultures, over centuries, have internalized a belief that people should have dominion over nature (Cronon 1991). Clearly, humans are anthropocentric. A new enlightenment about our place in nature is just beginning to emerge in the literature of modern science. In other scientific traditions, the human–nature relations have been seen in the form of interdependence. However, these traditions have a very small scale of operations in today's world.

3.5 Society in 2010

As the twenty-first century unfolds, it becomes evident that humanity is off-course to achieving the sustainable, peaceful and equitable future it may have desired. Failing and misdirected governance played a major part in this. Although from the 1990s onwards, the concept of "sustainable development" started to emerge as a global objective, nearly 20 years later it has not taken root in our way of life. Towards the end of the first decade of the twenty-first century, two particular and large obstacles on the path to a more sustainable and prosperous future are evident.

First, many societies and global relations are plagued by vast disparities in power. High levels of inequality are evident even in well-developed economies, and in many countries large numbers of people are economically, socially or politically excluded. This has been fostered by resistance from established power blocks, including those at national and global levels, to establishing inclusive, transparent processes. Further, globalizing processes have begun to severely undercut democracy, well-being and ecosystem integrity. The inequality in global power effectively makes democratic decisions within one nation-state undemocratic when they affect the rights of citizens of other nation-states. In many nations, executive multilateralism – weak national accountability mechanisms for foreign policy – and the relatively weak capacity and legitimacy (in some dimensions) of civil society actors, both NGOs and corporate associations, keep the vast majority of people on the fringes of decision-making.

Second, many of the choices about how to distribute and govern resources, including natural resources, are driven by an impetus to act in one's self-interest. The "tyranny of the actual over the ideal" persists. This is characterized by the reluctance of states to collaborate on key issues underlying unsustainable production and consumption systems, including energy. Even for issues with obvious global public goods attributes, like promoting energy efficiency and the transfer of renewable energy technologies, only weak collaboration efforts are made in sub-groupings of the more powerful countries such as the G8, OECD or smaller coalitions of the willing, while true multilateral efforts remain elusive. The stronghold of self-interest positions is also starkly apparent in the arena of international environmental law-making. At the end of the twentieth and beginning of the twenty-first centuries, much effort has gone into developing a legal framework for addressing urgent ecosystem threats. Although in most cases the agreements were very ambitious in general terms, they lack sufficient follow-up mechanisms, whether through enforcement, positive incentives or capacity building. Indeed, many states refuse to accept mandatory reporting of multilateral environmental agreement implementation, which is a minimum requirement for evaluating any change of behaviour in response to the agreements. At the national level, corruption and "elite capture" feeds distribution systems that keep the poor impoverished. In many developing and emerging economies, corporations are able to produce with little concern for the social or environmental impacts, while in developed countries many opt to export these adverse costs of production to nations other than their own.

A weak culture of responsibility, not only for nature but also for others, underpins the grave situation the planet and humanity face at the beginning of the century.

3.6 Governance

The global vision of development was defined by ideas of justice and peace, and a society in which all people lived in dignity and with respect for others and nature. The normative framework was far-reaching, touching on diverse aspects of life. In 1992, the United Nations Rio Declaration on Environment and Development[1] set a more holistic framework for dealing with development issues, but in general was poorly implemented. In 1995, this was followed by the UN Declaration on the Right to Social Development[2] and the Beijing Declaration (on women),[3] but 20 years later more than one billion people are still living on less than $1 per day, and gender inequality continues. In the decade that followed, further elaboration of these ideas took place and various commitments were made to realizing these ideals. Nevertheless, efforts continued to fall short.

Ever since publishing the "**Our Common Future**" report from the United Nations World Commission on Environment and Development, global discussions on challenges facing humans on Earth and setting goals for desirable futures have been growing. "Agenda 21" resulting from the 1992 Rio de Janeiro UNCED Earth Summit on Environment and Development, the "Joint Programme of Implementation" from the World Summit on Sustainable Development, the Kyoto Protocol of the UN Framework Convention on Climate Change, and the Millennium Development Goals show the human desire to find strategic pathways towards desirable futures.

Are these global diplomatic efforts making human life on Earth more desirable? As often argued on the premises of scientific research and well-planned strategies at the highest international levels, why has this desirable future become more and more challenging during the past 40 years of global sustainability visioning?

Uchita de Zoysa

Values that undercut these objectives (consumerism, short-sightedness and individual and national self-interest) have come to dominate the ways in which we organize our societies and are key factors in this failure. By 2008 – amidst a growing financial crisis – the United Nations warned that it was becoming more difficult to achieve the Millennium Development Goals.[4] One consequence is that 75 million more people, in addition to the previous 850 million, are victims of hunger and face acute hunger due to surging food prices (FAO 2008). This has the knock-on effect of reducing future options by increasing child malnutrition and reducing educational success. Global hunger threatens to spiral out of control even though the world produces more than enough food to feed everyone, highlighting the enormous threat that inequality and injustice pose to well-being. Human-driven destruction of ecosystems and the loss of biodiversity are rapidly reducing human opportunities. Even as the devastating realities of climate change become evident, nations struggle to find a shared, but differentiated, path to addressing climate change.

Government and markets have been the dominant institutions for dealing with climate change and are probably the existing institutions with the greatest impacts on energy use and the development of energy alternatives. Generally speaking, governments have five basic policy options for addressing climate change: prescription (e.g. explicit regulations concerning energy technologies, such as banning coal-fired power-plants), penalties (e.g. taxes or fines for emitting carbon), payments (e.g. subsidies for new energy technologies), property rights (e.g. cap-and-trade systems) and persuasion (e.g. educating people to reduce carbon emissions).

In practice, the world's most powerful nations – and the United States in particular – have deep faith in the ability of the market to solve resource problems. Market prices in theory serve as a fulcrum that balances supply (costs) with demand (benefits); what is possible with what is desirable. When a resource grows scarce, prices increase, consumers reduce demand and suppliers increase production or develop new substitutes. Many government policies therefore involve the "market" mechanisms of penalties, payments and property rights. For those who believe that the market can solve resource problems, global climate change is considered to be the result of correctable market failures. Nobody currently owns the waste absorption capacity for CO_2, which results in over-use by everyone – the inappropriately labelled "tragedy of the commons" (Hardin 1968, 1998) which is really the tragedy of open access (Ostrom 1990; Bromley 1992). To let the market work its magic, governments can create tradable property rights (a cap-and-trade system). Another diagnosis is that climate change is the result of negative externalities of fossil fuel use. In this case, carbon taxes force economic actors to internalize these external costs into their decisions, leading to socially optimal emission levels. Yet a third diagnosis is that global climate stability is a public good, which markets fail to provide. In this case, subsidies for developing and adopting technologies that reduce carbon emissions are appropriate. Unfortunately, most existing government subsidies perversely encourage fossil fuel use (Myers and Kent 2001). With payments and penalties, prices determine emission levels, while with cap-and-trade, allowed emissions levels determine prices (Daly 1997). National governments, primarily European, are experimenting with cap-and-trade, green taxes, feed-in tariffs (a guaranteed higher-than-market price for alternative energy, i.e. payments) and so on. The Kyoto Protocol is an intergovernmental global institution seeking to implement cap-and-trade.

> My area of work is energy for sustainable development. Energy system characteristics is closely linked to many of the challenges: two billion people have no access to modern forms of energy, without which there is no way out of poverty; fossil fuels contribute some two-thirds of greenhouse gases; peak oil is an area of concern that requires some form of action; population and economic growth adds to the demand for energy services. Energy systems for the future must be designed to meet all of these challenges *simultaneously and adequately*.
>
> Thomas B. Johansson

3.7 The global economy

"Business as usual" is a major challenge to the economic enterprise of our planet. Many of our current mechanisms, processes and institutions prevent equitable distribution of assets that contribute to human well-being. Globalization is completely transforming our mutual dependencies and the economic transactions that bind us. At the same time, the mechanisms and results of globalization are experienced by the vast majority of humanity as some new variant on imperialism.

> One of the big questions facing countries around the world, especially rich countries, is the place of economic growth in the panoply of societal and government objectives. It is only since the early 1950s that the deliberate pursuit of economic growth has been an explicit policy of governments in western countries. At first it was adopted as a requirement for full employment, but it soon became an important policy objective in its own right. More than that, by 1960 economic growth had become the overarching economic policy objective in many, many countries, virtually synonymous with progress, and who could question that?
>
> [...]
>
> While much is left to complete the picture, it is worth knowing that continued economic growth is not essential for rich countries to prosper. This is not to say that zero economic growth should be adopted as a policy objective. Rather that we should not be afraid to implement policies or to change how we work and live because it may be bad for economic growth.
>
> Peter Victor

The financial crisis that started in 2008 presented an opportunity – and even the necessity – to rethink our economic paradigm. For example, the ongoing, accelerating, negative consequences of our focus on "growth at all costs" beg for a change. We have the opportunity now to rethink many things, including policies, incentives, disincentives, technological and societal solutions. We need to inform, educate and empower our citizens, our markets and our institutions to execute best solutions locally, regionally and globally.

3.8 The role of belief and values

Belief and value systems manifested in a variety of socio-cultural institutions have invariably played a vital role in maintaining a harmonious relationship between societies and their environment in traditional societies. In large parts of the world this relationship still exists based on the firm conviction that the health of the Earth and all its natural endowments is crucial to human survival. Injudicious use of nature and its bounties boomerang back on society rather quickly through food scarcity, not buffered by distant markets and trade relations. Thus, over time humans devised ways to regulate the sustainable use of their environment through a variety of cultural forms, taboos, beliefs and value systems. Here the distinction between socio-cultural institutions, the economy and the environment remains blurred. Many of the rituals and ceremonies had a direct function to regulate the use of resources according to seasons and to confer sanctions in case of their misuse. Equity in terms of resource distribution took precedence over all forms of accumulation and wealth creation. Furthermore, rituals also served to reproduce the society and its power structures, and re-emphasize the role and importance of elders, thereby providing social stability and guided change. In times of crisis or difficult periods, cooperation and reciprocity was emphasized and ritualized. In such a system, it has been observed, only about one-quarter of the total available working time went into activities that had a direct bearing on nature in terms of resource extraction, while a large part went into leisure, festivities and rituals that kept the various components of the system in place. In essence, all vital components of the socio-ecological system are inextricably linked with positive and negative feedback loops to provide robustness, stability and guided change. When the regulation of the use of resources failed, however, systems collapsed (see, for example, Diamond 2005).

> Our minority world lifestyles are directly dependent on unsustainable material and energy consumption that threatens the otherwise self-regulating, complex life support systems of this planet.... We have created a system where I can live a perfectly good life by the standards of my immediate community, which is unsustainable and deeply unethical on the global scale.
>
> Stefan Kaufman

It is widely acknowledged that traditional systems together with their adaptive belief systems and values are under increasing threat. This process began much before the industrial revolution in Europe, with western science strongly influenced by notions of empiricism and rationalism of Descartes and Bacon. In order to reveal the secrets of nature it became imperative to distance oneself from nature and to place it under duress. In doing so, nature reveals itself in a way that allows humans to subdue it and exploit it for human development. With the industrial revolution the process of ruthless exploitation of nature was accelerated. An important prerequisite of this process was doing away with values and belief systems that played a crucial role in regulating natural resources. Thus, the present dominant global cultural influence has lost its awareness of the need to co-evolve with the ecosystems they live in, and it is this loss that is producing both the presumption of superiority and the dire need to change.

Religious organizations have been and are a dominant force in the world (see, for example, UNEP 2000). In 2000, over five billion people were adherents to a major world religion, with four billion belonging to Christianity, Islam or Hinduism, the three largest world religions. Sometimes these organizations can be an obstacle to sustainability – such as when they create barriers to interfaith acceptance or hinder effective family planning and prevention of sexually transmitted diseases. Yet these institutions can also be a powerful initiator of positive social transformation. Throughout history, when these institutions have flexed their moral muscles, the world has moved. During the 1700s and 1800s, when slavery was still seen as morally acceptable in many parts of the world, many Christian denominations took a leadership role in denouncing it and working towards the abolition of slavery. More recently, the Christian church took a strong role in the US civil rights efforts, and even more recently in stopping the inappropriate marketing of baby formula in places where a lack of access to clean water (mixed with formula) was leading to significant increases in infant mortality. Religious institutions have also taken an initiative to address climate change in the form of the development of global seven-year action plans.[5]

Although religious faith has been an important determinant of people's values, advertising has become much more important in the secularized world of the past century. In the early twenty-first century, over $500 billion was spent each year on advertising, predominantly in the richest countries of the world, but rapidly spreading with globalization. This advertising strengthens a culture of over-consumption of processed food, cars, electronics and other very profitable items that businesses sell (and people buy). But in the process of selling stuff, these advertisements (made more persuasive by some very sophisticated psychological tools) change humanity's value systems, encouraging us to view ourselves as successful and happy when we have access to these socially celebrated goods, and failures if we do not. Unfortunately, this cultural norm has a series of negative side-effects: often fuelling not just ecological damage, but unhealthy lifestyles and an increase in sickness and social isolation as well.

3.9 Quality of life

In 2010 the tacit driving goal for human society is ever-increasing GNP. While, perhaps in theory, increasing GNP need not require an ever greater flow of energy and raw materials and generate an ever-increasing flow of waste, in practice it always has (Smil 2003). The net result is the degradation of Earth's ecosystems and the life-sustaining services they generate. In other words, the current economic goals conflict with what is biophysically possible. Even if we accept the existence of biophysical limitations, we may do nothing to address them if they conflict with our goals, especially if we perceive the risks to be spatially or temporally distant. When US Republicans defeated the Lieberman–Warner Climate Security Act in 2008, their argument was that it would hurt the economy (Craig 2008). Of those Americans who believe climate change is human-caused, 54 per cent believe that addressing it would require major sacrifice, and overall Americans say that they are less willing than citizens of other countries to make sacrifices (Pew Center 2008).

There was considerable discussion towards the end of the first decade of the twenty-first century about the need for growth of GDP and about the need for other measures of societal progress (see, for example, www.beyond-gdp.eu). Abundant empirical evidence from countries such as Costa Rica and Cuba (UNDP 2008) and the region of Kerala in India (Mckibben 1995) confirm that society can attain the same expectancy of healthy years of life and education levels as western Europe, but with much, much lower levels of GDP. Further evidence suggests that beyond a certain threshold, increasing GDP – a proxy for consumption of market goods and services – does little or nothing to improve quality of life as measured by health, education and happiness (Easterlin 1974; Max-Neef 1991; Lane 2000; Diener and Biswas-Diener 2002; Costanza *et al.* 2007). In fact, in several wealthier countries, happiness levels appear to be declining while stress, anxiety and depression are on the rise (Lane 2000; Layard 2005). In the United States, per capita consumption has more than doubled since 1969. At the same time, poverty rates have risen, and the genuine progress indicator and subjective levels of well-being have fallen (Farley 2008).

3.10 Looking forward from 2010

The human species has advanced at a dizzying pace through the rapid evolution of the brain and its mind, leading to advanced powers of cognition, problem-solving and adaptation. It is this phenomenal development that has both got humans into a difficult situation, but can also get them out of it. Two fundamental characteristics of human behaviour are communication and memory, which together allow us to remember, respond to and learn from past events.

> We either continue our course with unknown consequences shifting the burden to future generations or we actively pursue a path towards mid-century, where civilization can look back upon the first half of

the twenty-first century knowing that we have turned the tide and are well on our way to recovering a healthy biosphere and well-being for our species.

John Katzenberger

The global village that we now inhabit is bursting at the seams, and unlike the isolated cultures of the past, which may have imploded when they reached their environmental limits, our societal response must now be global. Many of the necessary advances in human society, whether technological or behavioural, are without precedent. Global consciousness, international equity, enlightened and restrained lifestyles and non-monetary drivers of development present new, exciting, uncharted waters.

While there are examples of thoughtful action in response to environmental challenges – such as the soil conservation policies and statutes to mitigate erosion following the American dust bowl of the 1930s and the Montreal Protocol limiting the production and use of ozone-depleting substances after the ozone hole was confirmed in the 1980s – global consensus and policies have generally been slow and ineffective with regard to implementation of climate change policy (e.g. the Kyoto Protocol). It appears that a broad societal dialogue is needed in order to discuss possible pathways forward.

The analysis of the situation in 2010 points to a need to inform, educate and empower our citizens, our markets and our institutions to execute best solutions locally, regionally and globally. Furthermore, there is a need to recover an indigenous cultural outlook and praxis held prior to participating in colonial expansion.

The analysis also suggests that if we fail to understand that energy is required for all economic production and that the capacity of our atmosphere to absorb waste from fossil fuels is limited, then we will assume limitless economic growth is possible, as both Nordhaus (2007) and Stern (2006) do in their models. Instead, we must recognize biophysical limits and change our goals from economic growth to qualitative improvements in human life, such as education, life expectancy, social relations and the myriad benefits generated by healthy ecosystems.

Much of the discussion of the state of the planet in 2010 points to the central importance of good governance in numerous areas, from managing the quality and quantity of water, a most basic resource, to the creation of cities of inclusion and acceptance. It also underlines the fact that the world is a diverse place. From coral reefs to fisheries to forest management to agriculture, rangeland, rural and urban environments, there are no simple solutions to fit each unique problem. Prior to 2010, society has responded through partial solutions to specific problems without a systematic and careful consideration of impacts, or effectiveness with respect to an integrative approach to the innate coupling between human and environmental systems.

The nature and magnitude of the challenges ahead is enormous, as indicated by many of the points raised in this chapter. Providing all of the global population with clean water, sanitation, energy, shelter and food – some of the UN Millennium Development Goals – while respecting planetary limits, such as not exceeding "a dangerous level" of climate change, as agreed in the UN Framework Convention on Climate Change, will require great efforts worldwide. We first describe our normative vision for the planet in 2050 in the next chapter, before addressing the ways that these challenges could be met.

4 The vision of 2050

4.1 Introduction

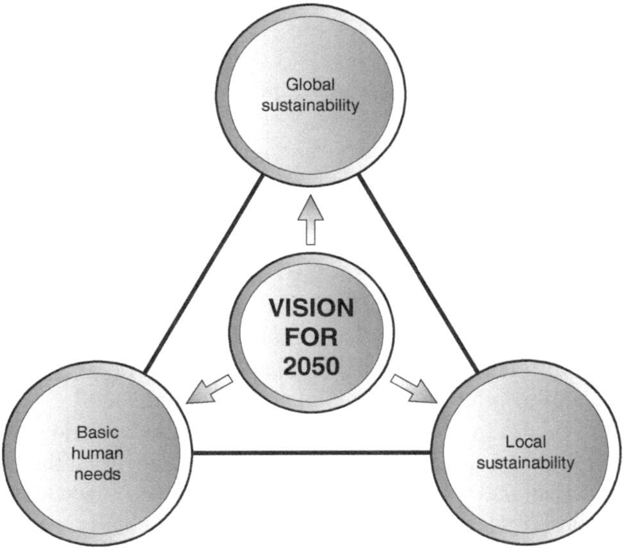

Figure 4.1 Elements of the 2050 vision (source: Workshop participants).

As shown in Figure 4.1, the vision for 2050 relates to three important elements:

- satisfying **individual basic needs** (e.g. food, water, shelter requirements);
- achieving **local and regional sustainability**;
- bearing in mind the **global sustainability** requirements.

4.2 The broad vision of societal development

In 2050, the following criteria must be met:

- All people have their basic human needs fulfilled, have equitable access to opportunities, and are able to pursue sustainable livelihoods and achieve well-being;
- People experience a sense of place and belonging in nature and a sense of purpose and happiness in the world. They have learned to value connections to their inner life through emotions and spirituality and base their personal relationships on mutual trust, care and compassion;
- Human societies value expressions of global solidarity and peace, respect for cultural diversity, peaceful conflict resolution, cooperation, trust and responsibility. Governance processes are characterized by transparency, accountability, openness, inclusiveness and participation, and decision-making at all levels is shaped by a sense of interdependence and unity in relationships within a single human family and the Earth system as a whole;
- The economic system is built on a paradigm that focuses on development rather than growth and is regulated from the global to the local level based on principles of justice.

The different components of well-being of the planet/Earth system achieved in 2050 can be captured in the form of a wheel (see Figure 4.2), where ecosystem health is the centre and foundation for human and societal well-being at different levels (local, national, global). In the following, we describe these tremendous achievements in more detail.

The wheel articulates the components of a health and well-being system in which human health is deeply embedded in and dependent on the health of ecosystems across the world. Without all five spokes, the wheel will either fall apart or wobble. Each spoke in the wheel is actually comprised of threads that connect formal and informal institutional aspects of that spoke, and when they run the length of the spoke, they connect individuals and communities to the ecosystem. The more connections that run the length of the spoke that exist, the stronger is the bond between the ecosystem and other existing systems. But if these threads are not woven through, just like a damaged tyre, they will unwind. Many interventions that focused on only one part of the spoke – either those closer to the ecosystem or those closer to individuals – failed; only when they were woven tightly together did they become effective.

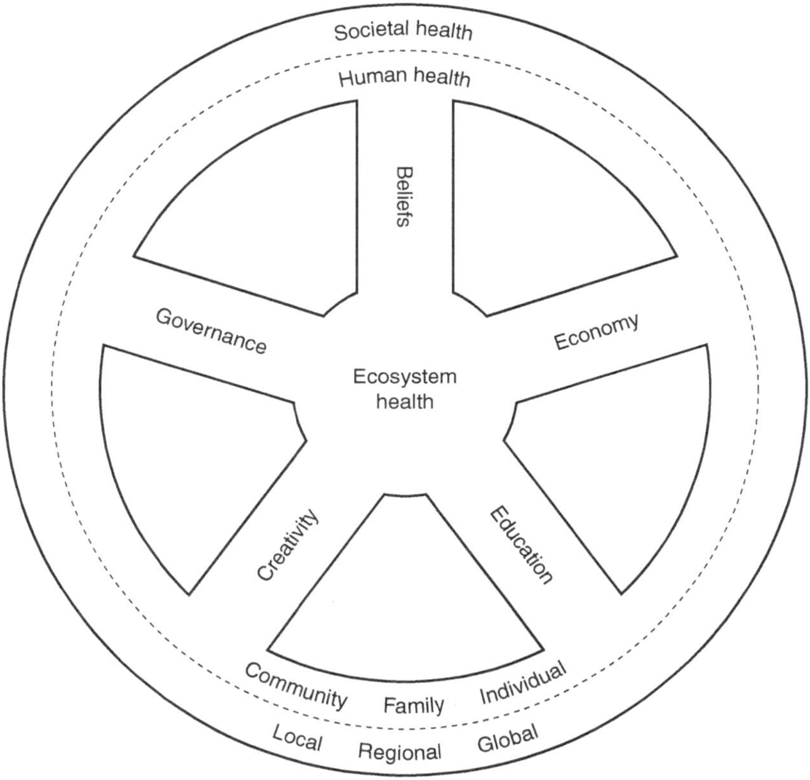

Figure 4.2 The wheel (source: Workshop participants).

4.2.1 Sustainable livelihoods and human well-being

In 2050 all people – in all continents and countries – have achieved a level of well-being that allows them all to pursue sustainable livelihoods along economic, social and political paths that support them in achieving meaningful, creative and fulfilling lives. This means that hunger and poverty are cast aside; that all people are able to attain essential needs such as food, clothing, housing, healthcare, education, energy and clean water and sanitation; that under-fives no longer die from the preventable diseases that were rife at the start of the twenty-first century; and that the elderly do not die from cold or heat exposure. As a result, all people, including women and other previously marginalized people, participate fully in society and have equitable opportunities. Human health is no longer considered outside of the health of the ecosystems in which people live, and ecological awareness is an intrinsic part of the education system. People are able to respond effectively to social and environmental hazards, and societies care and provide for the most vulnerable amongst them.

> A deeper understanding of society–nature interactions has been achieved in favour of local, small-scale decentralized development models of production and consumption that balance (local) human and natural capital with respect to the social and cultural context. At the same time a reduction of the dependency on fossil fuels has been accompanied by a search and shift to alternate sources.
>
> Simron Jit Singh

4.2.2 Well-being within and beyond

People in 2050 have achieved a deeper sense of place and belonging as part of one human race on this planet. The human relationship with nature, or rather the human position within nature, has become more encompassing. It has become more logical to appreciate all living things, as we have increasingly realized over the past decades that we are but a part of the greater whole and that valuing the natural world is akin to valuing ourselves. This underpins the strides that we have taken in the pursuit of a prospering civilization, which get inspiration from and celebrate the awe-inspiring complexity, diversity and grandeur of the natural world. In our world this cultural shift from the previous century manifests itself in many ways – through spirituality and self-reflection; deep appreciation for creativity, genius and the arts; the strength of our community relationships; and a sense of purpose in the world.

> Humanity has embraced diversity as a highly desirable feature of human systems, actively nurturing the rich diverse cultural heritage and actively drawing on it in designing institutions for governance of the Earth system.
>
> Sylvia Karlsson-Vinkhuyzen

Society now respects universal values and promotes human dignity to a much greater degree than in previous periods of human history. Human relationships are based on mutual trust, care and compassion and our uniquely diverse cultural heritages are valued and deeply appreciated. People give time and space to thoughtful care in building and maintaining human relationships and respectful relations with the natural world around us.

It's 2050, Africa prospers and her people have secured sustainable livelihoods; hunger and the lack of opportunity no longer persist. In Southeast Asia, the benefits of economic growth are shared more equitably. Worldwide, a new understanding of community has emerged in which all are "citizens".... The right to development along with rights to water and food are widely recognized and implementation efforts are supported by developed and developing governments, and other duty-bearers and rights-holders.

Jennifer Mohamed-Katerere

These dimensions of well-being – beyond the mere material – touch people's souls much like the beauty of nature does, because they create emotional reactions in us that we cannot truly all understand or communicate, but that we can obviously feel. Manifestations of this cultural shift include the time people spend in their gardens (urban or rural), in the parks and nature which surround them. No child misses out on learning the value and practice of agriculture and the numerous ways in which ecosystems nurture us. Villages and cities have given much greater attention to integrating beauty and nature, and the centuries-old experiences in managing local natural resources such as lakes, forests and grazing land jointly by communities have been strengthened and multiplied across the world – also strengthening relationships and connectedness with nature. In valuing emotions, spirituality and those connections to our inner soul, humans are now acknowledging – joyfully – that the inexplicable is not only what makes us human, but what makes us part of the Earth system.

4.2.3 Governance based on values

The political dimension of our societies is now aware of the importance of relationships between its two main agents: the institutions and the people, which in turn also interact with each other. Within these relationships, peace, non-violence, participative democracy, social justice and global equity are the defining values of societies. The dignity of people and integrity of nature underpin decision-making at all levels. We have reached far to make transparency, accountability, openness, inclusiveness and participation the defining characteristics of governance processes so that policies that come out of these processes are effective and coherent across levels and sectors. The decision-making process itself is characterized by mutual learning (the role of knowledge, science and education is therefore considerable), cooperation and dialogue. This builds unity and ensures peaceful relations at all levels that provide the foundation for the prosperity of humankind, including and beyond material dimensions. There is a wide acknowledgement of interdependence – among nations, people and communities, as well as between them and nature – which in all domains (economy, politics, social) shapes choices and decision-making at all levels. The principle of global equity and ecological interdependence is so integrated that no longer does one section of our world make choices that harm the well-being of others or compromise ecological sustainability.

> I envision a comprehensive effort to build a long-term (meaning multi-hundred-year) movement, using the strategies that have worked since adherents of Buddhism, Christianity and Islam first figured out how to spread their philosophies across the globe. Specifically, these adherents established a common, awe-inspiring philosophy and then *organized* to spread this philosophy. They set up meeting houses, social service providers that help people in need while teaching their way of living (models include the Salvation Army, Catholic schools, Madrasas and Jewish hospitals, to name a few modern variations). By 2050, environmentalists have done the same, but this time the philosophy was based on a belief that we are dependent on and part of a living Earth system, and that meant a redefinition of how we should live on this Earth.
>
> Erik Assadourian

The personal dimension of the political has become rooted in a sense of community and citizenship in overlapping spheres from the local to the global. All are recognized as citizens, and previously marginalized groups such as women and indigenous peoples have become valued participants in all spheres of society. Citizens also take personal responsibility and participate actively in democratic processes across levels. A widespread orientation towards the service of others provides the foundation for this level of responsibility and the prosperous societies we have built. Diversity of our human and environmental systems is now understood in a context of unity, and both are cherished as fundamental components of society. A significant portion of government spending is put into enhancing and cultivating biodiversity, including sustained investment in mutually beneficial management of ecosystem health and employment opportunities for communities to undertake these measures.

4.2.4 An economic system based on justice

> The global financial collapse of 2008 instructed humanity that the unconstrained, free-market hyper-consumptive economies of the past were not sustainable in a finite world. The mantra, that economic

growth of 4–6 per cent per year was an essential part of all national planning, could not continue indefinitely. The dominance that global economics had on international governance in the twentieth century gave way to a more cooperative goal of mutual security and shared prosperity, allowing dramatic decreases in national military expenditures and increased investment in human resources.

Steve Running

The paradigm underlying the economic system has shifted from "growth"-oriented to "development"-oriented. The new economic system values all the diverse kinds of capital and wealth (social, human, cultural, ecological, social, infrastructural/physical, etc.) as the core assets of society. The indicators that are used to track progress in development have been greatly expanded over the past decades and now include qualitative and quantitative measures of all these types of capital and their interdependencies. On a macro scale, we have strong global financial institutions governed by the principles of accountability and democracy, which have developed global rules that regulate international trade and financial flows in ways that discourage speculation and encourage more equitable sharing of resources. Now the various economic sectors value socio-environmentally certified products, and incentives for investment opportunities in such sustainable products and processes are offered globally. Local and regional markets and cooperative structures are strengthened as an answer to the high-carbon character of the former globalization model. International trade still plays an important role, but is concentrated on services and goods for which the high transport emissions are mitigated by lower environmental impacts in production and which have a greater focus on benefits to local communities where the services and goods are produced. The economy promotes appropriate technology (in energy, sanitation, communication, food production, etc.), including the development of funding mechanisms for products and processes with much longer lifecycles or full recyclability.

4.2.5 Far but further still

Underlying the achievement of human individual and collective well-being that has been realized by the mid-2000s and described above is a widespread appreciation of our shared destiny – shared with all people and with nature. Solidarity, inclusion, peace and respect for nature are defining values, and this manifests itself in the way that humanity now lives consciously and deliberately within the Earth system's biogeophysical cycles and the means and limitations these provide. This does not mean, however, that we stand without challenges. Debts in terms of environmental degradation and human suffering incurred over past decades and centuries still demand considerable remedial action. And with every achievement, we identify higher aspirations to strive for in the wheel of well-being. Humanity's continuing journey, with and as part of this planet, has barely begun. But we are grateful that we have now at least ensured that we have not cut the journey short through our own folly, and instead put in place the fundamental elements for the journey to be one where many aspects of human suffering of the past have been virtually eliminated, and one which can last a very, very long time into the future.

By 2050 the dominant practices, rules and assumptions of the European Union have changed dramatically from those that were in place in 2008. The prices for energy, resources and land truly reflect the value of ecosystem goods and services. Society measures its progress not in terms of economic wealth, but in terms of well-being of people and the environment both in Europe and worldwide. Paid work is distributed, so that all people who want a job have a job, while work that was previously unpaid (caring for the young, the old and the sick, voluntary work to support environment and society) is paid through a guaranteed basic income for all citizens.

Jill Jäger

4.3 Energy in 2050

Our vision is that social and engineering barriers to a revolutionary transformation to sustainable energy have been overcome by 2050. Given the geophysical realities of limitations to resource supply and to the absorption of waste, industrial society has had to reinvent itself, not only with revolutionary new green energy on a massive scale, but with industrial ecologies recycling scarce materials.

In 2050 the energy system:

- **Is carbon-neutral (or almost so)** – revolutionary transformations of our fossil-fuel-based system of the twentieth century to one that is carbon-neutral with little net CO_2 emitted have taken place by mid-century, in parallel with aggressive improvements in energy efficiency. These huge emission reductions leave most of the Earth's coal in the ground or sequester its emitted CO_2.

- **Provides universal affordable access to modern forms of energy at the levels ensuring a high level of human well-being** – an equitable world in 2050, where most of the development targets articulated in 2010 have been achieved, implies universal access to modern energy services to meet basic needs like heating, cooling, lighting, mobility and communications. A priority was to provide energy services to the roughly two billion people who, in 2010, had no access to electricity or clean cooking fuels – a systemic cause of poverty, poor health and other social problems. Achieving carbon-neutrality with equity has required new technology and lifestyles that have broken the correlation between per capita carbon emissions and per capita GDP.

New system-enabling technologies are smart electric grids, distributed-generation, long-distance high-voltage DC transmission and massive energy storage.

- **Does not have lasting, widespread negative effects on or risks for health, environment and security.**
- **Is on a sustainable pathway** – the energy system of 2050 (1) does not face risks of major disruptions/collapse on a time-scale of at least several centuries; and (2) is capable of adapting to changing natural, technological, demographic and social conditions.

Box 4.1 Carbon-neutral technologies that have made a difference

For electricity

Wind
Marine renewables
Solar thermal photovoltaics
Coal gasifier multicycle plants with carbon capture and storage
Centralized solar power

For end-use-demand reduction

Super-insulated buildings
Efficient lighting
Urban transformation

For emission-free vehicles and infrastructures

Biofuel, hybrid, electric and hydrogen-propulsion engines

4.4 The environment in 2050

4.4.1 Water

In 2050 the challenge of balancing the use of freshwater for humans and nature has been met as a result of the introduction in the early twenty-first century of an integrated land and water governance and management approach that incorporated the role of water to sustain both terrestrial and aquatic ecosystem functions and services. Now, on the whole, freshwater resources are safeguarded to provide direct human well-being (water for food, energy, fibre and timber), as well as water for terrestrial and aquatic ecosystem functions, goods and services.

This approach, including both green and blue water resources, enabled a strategic move towards safeguarding freshwater for carbon sinks, biological diversity, food, industry and domestic water supply. The challenges in 2050 remain tremendous. With a world population of nine billion people, and the successful achievement of the UN Millennium Development Goals (the complete eradication of malnourishment, which amounted to one billion people in 2008), the global freshwater use for food production on the planet has increased by a staggering 3000 km^3/yr in 2050, up from approximately 7000 km^3/yr at the end of the twenty-first century. However, this is seen as a major global success, as all predictions in the late twentieth century pointed to an increase of at least twice this magnitude. It is dramatic, given that food production was and remains the world's largest freshwater-consuming human activity.

4.4.2 Air quality

By 2050, the air of the cities of the world has generally improved, although there are still "hot spots" of poor air quality. The strategy to reduce emissions differed according to the region of the world, but they all required a limitation in fossil fuel production, changes in our modes of transportation and the development of new agricultural practices. Savannah fires and domestic burning, especially in the tropics, were specifically targeted.

A continuing source of air pollution is wildfires, which have global implications through their role in global biogeochemical cycles and potential climate system feedbacks. The release of soot particles into the atmosphere has been reduced by limiting the use of diesel cars, as well as the frequency of biomass burning events. Attention has been paid to reduction of a wide group of pollutants, including fine particles and their precursors, such as sulphur dioxide (a product of coal burning) and ammonia (a product of intensive agriculture). In addition, reductions have been made in the release of methane (another ozone precursor and a GHG) and nitrous oxide (a GHG linked to the use of nitrogen fertilizers).

4.4.3 Greenhouse gases

In addition to reducing emission of CO_2 from fossil fuel combustion, effective methods are now available to actively capture CO_2 from the atmosphere and store it in a stable form. The demonstration carbon capture and storage plants established in the first decade of the twenty-first century have proven robust enough, and negotiations are underway aimed at ensuring an annual capture of CO_2 from the atmosphere equalling the GHG effect arising from the global anthropogenic release of non-CO_2 GHGs. By essentially eliminating all anthropogenic influence on GHG accumulation in the atmosphere in this manner, it is predicted that atmospheric GHG concentrations will – in the centuries to come – eventually stabilize at pre-twenty-first century levels.

4.4.4 Oceans

In 2050 the oceans have received far more recognition in global environment governance, as they hold the key to long-term global warming (as a result of the inertia in the warming process) and the key to large parts of the carbon sequestration capacity. The acidification trend of the twentieth and early twenty-first centuries has been halted thanks to the reduction in CO_2 emissions from fossil fuel combustion. The pH levels in the world's oceans have now stabilized at levels required to avoid the risk of losing the ability of ocean species to absorb CO_2.

4.4.5 The land surface

On land, the few remaining rainforests, in the Amazon, the Congo, and in Southeast Asia, have been preserved as "global lungs" and carbon sinks. Forest systems worldwide are managed in order to increase their capacity to perform as global carbon sinks, as well as other management goals. Terrestrial ecosystems are maintained for biodiversity and as continued carbon sinks. The world's croplands have stabilized at around 15 per cent of Earth's land area (compared with about 11 per cent in 2000) in order to safeguard carbon sequestration and biodiversity of the world's terrestrial ecosystems. Furthermore, agricultural systems on Earth now have switched from being a source to being a global sink at a level of 1–2 Gt C/year, in particular through appropriate land management practices.

These appropriate land management practices have been implemented to provide sustainable levels of commodities, while simultaneously allowing the maximum climate regulation. There is no conflict regarding land used for biofuels, crops, pastures and forestry. In addition, landscape heterogeneity is maintained for production, biodiversity conservation and cultural reasons. Agriculture in 2050 has realized the ideals of sustainable farming, including optimized profitability of farming and reduced emissions. This has resulted in a balanced development between urban and rural areas and the establishment of equitable marketing channels.

4.4.6 Urban areas

In 2050, approximately two-thirds of the world's population lives in cities and towns (up from 50 per cent in 2010), which now cover about 8 per cent of the Earth's surface. From a global perspective there is a more balanced distribution of small, medium and large urban areas than found in 2010. The cities and towns of 2050 have creative, vibrant and diversified economies that contribute to the well-being of the whole population. The citizens enjoy better health than in the urban areas of 2010: food and water security, as well as personal security, have improved; and urban areas provide opportunities for living, caring, working, recreation and communication without having to travel large distances. Through integrated consideration of urban issues, the pressures of urban areas on the Earth system have been considerably reduced.

> Today the world is focused on the individual. In 2050, communities, relationships between people and personal responsibilities to society will be the key principle in the construction of our cities. The structure of the city will support stronger communities by bringing people together in more cohesive spaces, which are then tied together in larger and larger webs. This idea of a polycentric city and urban nodes renders the issue of mobility moot; people will have greater access to urban functions not only within their own nodes, but through convenient mass transportation connections between cities. Denser urban nodes will also eliminate the need for cars within these communities and provide more space for parks, urban agriculture and places for social interaction.
>
> The MESPOM student participants from IIEEE

4.4.7 Environmental governance in 2050

Multi-level governance systems ensure consistency between decisions at the local, regional and global levels. New governance structures at the global level are in place. These structures recognize that a prerequisite for a sustainable future for the Earth system is that human societies accept that they have a responsibility to perform as "guardians of the global element cycles". This responsibility includes having distributed Earth observatories and mechanisms and assessment abilities to enforce international agreements on the distributed carbon sink commitment among the nations of the world. The role of nature as a life-support system is a key component in all decision-making. All citizens of the planet are equipped with knowledge to act responsibly with respect to environmental challenges.

It is with incredulous looks in their faces that young students at Stockholm University, this spring term of 2050, listen to their professor of political sciences, describing the old, primitive, nation-state concept that prevailed just a generation earlier. This old structure crumbled to a dramatic end, after decades of growing mistrust between "nations", when it was found to be completely out of sync with the rapidly growing realization of the impacts on all world citizens of the Anthropocene.

Johan Rockström

By 2050, political, economic and environmental decision-making has, for several decades, been made with consideration of Earth system limits and functions, by assessing the potential impact of human activities on fundamental natural resources and the global elemental cycling. Uncertainties remain high, but an active precautionary principle is now in place and operational. Instead of trying to predict the future, humanity is now making all efforts to minimize risks of undesirable trajectories for present and future generations. This knowledge is used by humanity to stay on the safe side of planetary boundaries, i.e. "hard-wired" limits in the stratosphere, atmosphere, cryosphere and biosphere, which, if crossed, may cause undesirable state changes. As a global strategy, this has resulted in an explicit effort in 2050 to assure a high degree of redundancy and diversity in the sub-systems of the Earth system, including terrestrial, aquatic and marine biodiversity; concentration of GHGs in the atmosphere; and elemental cycles.

In order to minimize the risk of undesirable tipping points in the Earth system – that is, thresholds for which a crossing would lead to major and potentially irreversible changes – strategies are in place by 2050 that enable humanity to avoid pushing critical environmental parameters too close to the danger zone where tipping points cannot be excluded. In 2050 governance and management principles have abandoned the mechanistic optimization principles which prevailed in the twentieth and early twenty-first centuries. It has become abundantly apparent, from both science and practice, that the closely inter-connected human–nature relationships do not follow linear, predictable and controllable patterns. It was this assumption that formed the fundament for an optimization regime that dominated 50 years ago. We now recognize that socio-ecological systems exhibit a high degree of non-linear behaviour. This is true for all types of system, from local ecosystems such as lakes and forests, to regional systems such as the monsoon and rainforest climate systems, and global systems, such as the atmosphere, cryosphere and stratosphere.

4.5 2050: a stepping stone to the future

This vision for 2050 is ambitious, and achieving it depends a lot on significant changes in technology, lifestyles, values, behaviour and governance taking place between 2010 and 2050. The pathways to this ambitious endpoint are discussed in the next chapter. It is important to underline, however, that this is not a "final station". Because of huge time lags in the Earth system, this status report for the planet in 2050 must be seen as a "stepping stone" on the road towards a sustainable future.

The vision for 2050 is technologically and behaviourally optimistic. As we will see in the next chapter, it requires understanding the objective problem and applying appropriate political action and policies. Before dismissing it as "utopian", "unachievable" or "wishful thinking", we should perhaps briefly consider the "darker" alternative (see Box 4.2).

Box 4.2 The "darker" alternative for 2050

A business-as-usual strategy builds up the atmospheric CO_2 concentration from burning of oil, gas and coal to a *tipping point*, beyond which grounded ice-sheet disintegration on Greenland and Antarctica becomes irreversible, accompanied by massive continental land inundation and climatic migrants. Biofuel production has risen to meet the pressing energy demands. Zoos and labs retain genetics, but landscapes and marine environments have been completely transformed. Biological species extinctions mount rapidly as a result of climate change, along with massive coral reef and marine species diebacks from acidifying oceans.

This warming may accelerate from positive feedbacks, like degassing more CO_2 from higher soil bacterial respiration rates on a warmer Earth. Eventually a massive release of methane locked in polar clathrates occurs. A similar spike apparently happened 55 million years ago, the Palaeo-Eocene Thermal Maximum, linked to a carbon release to the atmosphere comparable to that in present-day global coal reserves (Pagini *et al.* 2006). The best that humans can expect might be retreat to polar refuges to await a restart of global civilization – the reset more difficult because fossil fuels for energy are depleted.

5 Pathways between 2010 and 2050

5.1 Introduction

Clearly, different strategies and actions were used in different contexts to ensure that the "vision of 2050" was achieved and that there was a move to "sustainability" – including efforts to eliminate the extremes of wealth and poverty and reach a convergence in consumption levels. It was seen as necessary to have a unifying element to all these strategies and actions in order to ensure that humanity reached its common goal, and this unifying element was a global, directional cultural shift. What happened to the environment between 2010 and 2050 depended very much on the decisions taken over that period by individuals and communities, and nations at the global level.

At the start of this endeavour in the early twenty-first century, a number of crucial questions were raised. How do we as humanity enable such a shift in the way we think, feel and act towards each other and the planet? How does one change culture? Is it a bottom-up process starting with individuals, or is it a top-down process from governments and leaders? Where do individuals draw inspiration for cultural change? Which are the most important sectors and actors to promote a cultural shift? Over time, the understanding took hold that at the root of such a culture change are values. Values influence how we interpret knowledge, feel concern for others and put such knowledge and feelings into action, and it became clear that the values that had come to dominate many societies, and particularly the most powerful societies, were aggravating or even driving the unsustainable development so obvious at the beginning of the twenty-first century; values that centred on self and national short-term interests and excluded the natural environment from human concern. Disengagement from such values and the reconnection with values existing in many cultures and faith traditions of the world, with a focus on consideration for all other living beings, became the key component in a cultural shift that was nurtured and started to manifest itself over the following decades.

It became clear that addressing symptoms alone – militarism, extreme poverty or climate change – was not a workable strategy to achieve the aspirations humanity had developed for a planet of well-being, but that what was needed was a directional cultural shift. This shift was supported through concerted action and change along five pathways, each of which provided a key avenue of change:

- governance;
- the organization and management of the global economy;
- knowledge and education;
- creativity; and
- value and belief systems.

In the following sections we describe the major strategies that were employed along each of these pathways over the past four decades to move towards the common goal. These pathways are illustrated in Box 5.1.

A guiding principle for the transition was equity. This demanded a reduction of resource use by rich industrialized countries and an increase of resource use in those countries that were still developing in 2010. This pattern has been referred to as "contraction and convergence" (see Figure 5.1). It was a pragmatic decision at least as much as a philosophical one – inequity was evidently problematic.

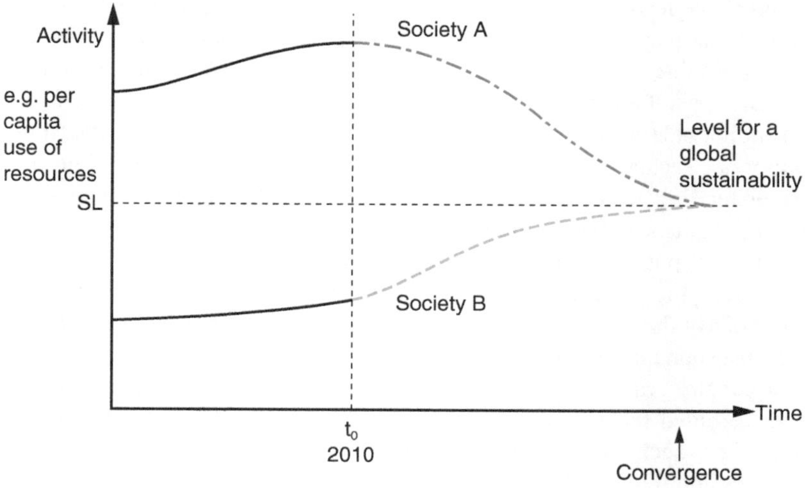

Figure 5.1 Sustainability demand for contraction and convergence (source: based on Jäger 2009).

Box 5.1 Pathways from 2010 to 2050

The wheel (see Chapter 4) articulates the components of a health and well-being system in which human health is deeply embedded in and dependent on the health of ecosystems across the world. The model has been helpful to identify and connect five major pathways which humanity employed to achieve health and well-being of society and individuals in the context of the ecosystems in which we live. Points of intervention along each pathway address the linkages between ecosystem health and public/societal health, either by strengthening the ties between the two, connecting disparate solutions that address only one aspect of overall health, or creating new linkages. Thus, there was room for all societies to strengthen the "spokes" of the wheel.

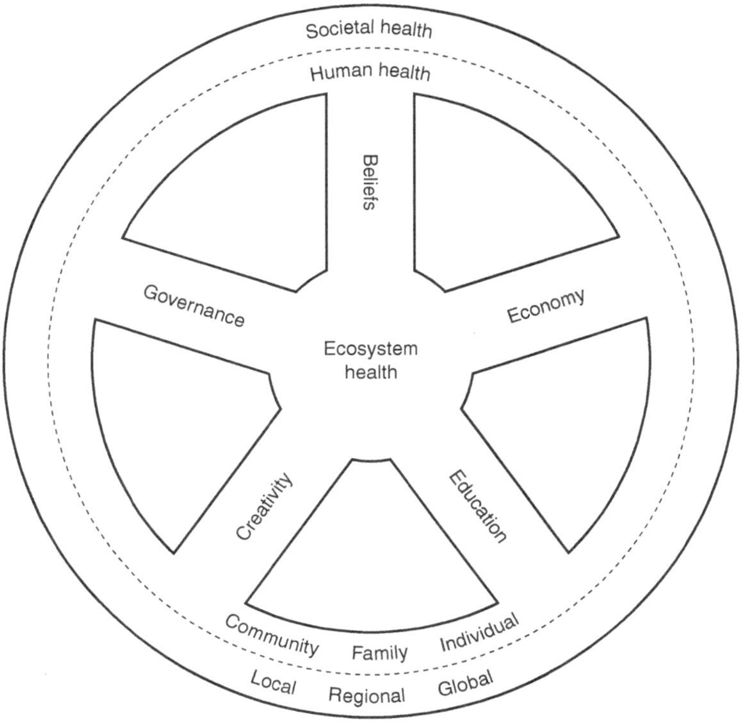

5.2 The pathways

5.2.1 The governance pathway

At the start of the transition period, growing numbers of nations, civil society organizations and individuals saw the need and urgency for reorienting governance systems so that these would more effectively support the realization of agreed objectives. A weak culture of responsibility, not only for nature but also for others, underpinned the grave situation the planet and humanity faced at the beginning of the century. It was evident that to address these challenges fundamental changes were necessary.

> Calls for greater decentralization of power clashed with the need for international efforts to deal with the global financial and biophysical crises. The result was a steady global movement towards governance based on the subsidiarity principle, with political decisions made by the least centralized competent authority, which for many serious problems required revitalized global governance structures.
>
> Joshua Farley

Seeds of a new beginning could be discerned, although they emerged in subtle and unobtrusive forms at first. Nevertheless, they turned out to be major opportunities for transitions. In the early part of the twenty-first century, it was evident that there was an emerging cosmopolitanism that envisaged the "possibility of global issues being addressed on the basis of new forms of democracy, derived from the universal rights of global citizens" (Chandler 2003). The existence of a shared framework based on the vision of how things ought to be, including a well-established human rights framework, international environmental law and policy and a minimum consensus on desired environment and human well-being objectives, reinforced the possibility for governance transitions.

A particularly powerful force for change was an increasingly radicalized public – no longer willing to sit on the edges of power and accept remotely made choices that threatened the social and environmental conditions for a sustainable future. By 2010 there was widespread disillusionment with politics at the national and international levels. Given new understanding of the urgency of the issues for people and planet, this did not lead to apathy as had previously been the case, but became a force for re-engagement in many different forms across all levels. Populations were not only more aware and more knowledgeable, but also demanded more actively to be recognized as citizens. There was extensive recognition of the links between actions that impacted the environment and the adverse social outcomes across time and space.

Seizing on these opportunities, societies, collectively and on their own, developed two strategies to ensure that governance – and its outcomes – were in harmony with core values. First, they developed systems (institutions and governance processes) that recognized citizenship, responsibility and equitable opportunity. Second, they reinforced cosmopolitanism by emphasizing commonalities and thus establishing the basis for collective interest and action. Consequently, citizenship took on a global, in addition to a national, meaning.

Recognizing and empowering people as citizens

Rights, knowledge and a strong emphasis on morality underpin the ability of people, individually and collectively, to assume the responsibility – the burden of ensuring the social good including the functionality and integrity of ecosystems – and opportunities of citizenship.

To develop this new sense of responsibility, systems that supported people and organizations, operating at different levels from the local to the global, to act more sustainably were adopted in all countries and within the global system. These systems included inclusive and interactive engagement with formal and informal educational institutions, which supported better understanding by both citizens and governments of challenges and possible solutions. At the national level, emphasis was placed not only on the responsibility and accountability of the state to its citizens, but also on individual, organizational and private sector responsibility and accountability. This approach recognized that nations are made up of individuals, with diverse views and values, and thus espouse multiple, sometimes contradictory behaviours. At the global level, the importance of both state actors and civil society groups was acknowledged; this included recognizing the role of new communities, including NGOs and new groups of civil society actors.

> Many social action groups working among the common people in all parts of the world, with input from independent professional experts and other collaborators, have succeeded in bringing desirable changes where governments have not. This provides a great opportunity for humanity.
>
> Jayanta Bandyopadhyay

At both the national and global level, a focus on the four Es – engage, empower (or enable), encourage and exemplify (see Figure 5.2) – facilitated this transition to greater assumption of responsibility.[1] This entailed:

- **engaging** populations with issues of morality and the implications of fundamental human rights, respect and dignity, justice and fairness, etc. for choices and vice versa;
- increasing efforts to **empower** (or enable) people to act responsibly by enlarging their environmental awareness and conscience;
- **encouraging** a diversity of voices and widening inclusion;
- **exemplifying** desired behaviour by encouraging institutions and public (and other) figures to lead by example, and by ensuring that policy offers consistent and compelling messages.

The empowerment efforts included environmental literacy campaigns, which – among other things – improved understanding of nature's limits and provided a better basis for making choices through more dialogues among science, civil society and policy-makers. This meant developing a new role for science as one – but not the only – provider of knowledge in societal processes. By improving access to education and health services and making social relations more inclusive and equitable, people individually and in communities had increased capabilities to make choices they have reason to value. This changed set of circumstances – a more just society, better understanding of its place in nature, new bonds of unity – increased the opportunity for effective social networks. Greater institutional accountability was ensured through better reporting, developing space for shadow reports in which dissenting views could be presented, independent verification of state reports and improved access to information. Exemplifying desired behaviour was supported, for example, by governments adopting green procurement policies that reinforced positive social attitudes and encouraged greater corporate social responsibility. Similarly, greening planning policy was encouraged, and supported increased uptake of green architectural practices by citizens. Ensuring consistent policy messages demanded ironing out contradictions between policies, removing perverse incentives and harmonizing approaches to governing and managing natural resources across sectors and temporal and spatial levels.

At the beginning of the twenty-first century many people, especially the poorest and most vulnerable, lacked the capabilities (such as education and good health) and the opportunities to live fulfilling lives. In essence, they were denied the rights of citizens, often in favour of elites, to enjoy an equitable share of resources, including environmental resources. As a first step towards rectifying this, societies dependent on natural resources recognized the links between realizing fundamental human rights, such as a right to food and water, and natural resource tenure. This not only helped secure livelihoods but also created the basis for long-term management

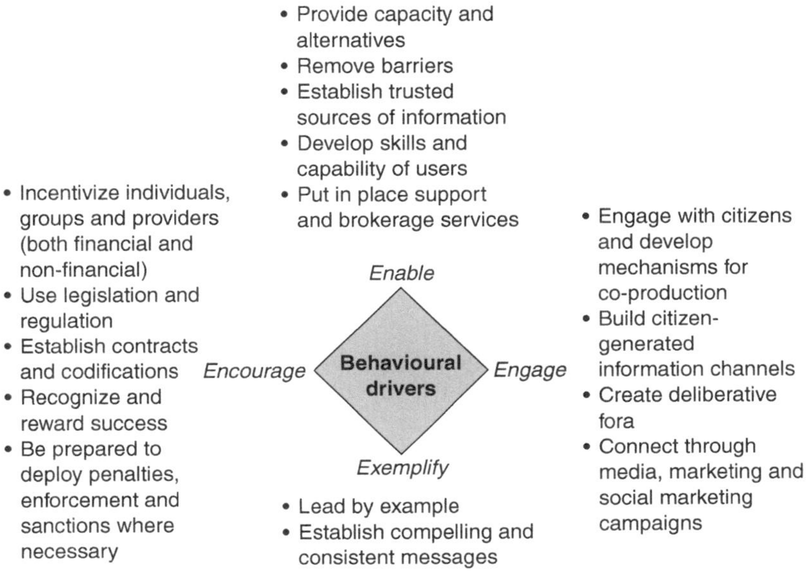

Figure 5.2 Policy interventions that change behaviour (source: UK Cabinet Office 2008).

perspectives and investment. Through systematic and concerted efforts to develop and strengthen inclusive policy and decision-making processes, society was able to decrease vulnerability and rid itself of the scourge of poverty. Inequity was significantly reduced. More inclusive policy processes created the conditions for a fair and equitable "balancing" of priorities and interests in making trade-offs. In addition, systems to ensure equitable sharing of costs and benefits of development choices were developed. This included addressing issues of environmental justice and ensuring that global responses did not undercut well-being.

Strengthening cosmopolitanism

By 2010, efforts to boost cosmopolitanism by institutionalizing possibilities for dialogue on commonalities and strengthening accountability were seriously underway. There was a fundamental shift away from adversarial systems to consensus-building through dialogue. This was achieved by active support for weak civil society actors, particularly those from the South and indigenous communities, to strengthen their ability to participate effectively at home and in international arenas. New categories of civil society were able to organize themselves more effectively to participate as collectives at the global level. Consequently, a fundamental transition occurred from a focus on the rights of the states as the underlying principle of international society to the rights of individual citizens. Human rights emerged as a benchmark for global and national choices.

More effective systems of accountability were also developed, transforming the relationship between state and citizen, as well as among states. This encompassed improving the reporting and follow-up systems for global commitments of all kinds, including in the financial realm, both through an international system of peer-review by states and civil society actors, and nationally through stronger engagement of parliaments in the formulation and implementation of policies. The seeds for a global "demos" had been sown and, after these seeds took root, engagements began around reforming the UN organs to include direct representation not only of states, but also of people's organizations.

The role of government had to be reinvented. In addition to the government's role in regulating and policing the private market economy, it has a significant role to play in expanding the "commons sector" that can propertize and manage non-marketed natural and social capital assets. It also has a major role to play as facilitator of societal development of a shared vision of what a sustainable and desirable future would look like. Strong democracy based on developing a shared vision was an essential prerequisite to building a sustainable and desirable future (Prugh *et al.* 1999). The vision for 2050 required a core set of principles for sustainable governance (Costanza *et al.* 1998) based on an integrated approach (across disciplines, stakeholder groups and generations) using the paradigm of "adaptive management", whereby policy-making is an iterative experiment acknowledging uncertainty, rather than a static "answer".

5.2.2 The global economy pathway

The driving goal for human society in 2010 was, as described in Chapter 3, to maximize consumption, measured by an ever-increasing GNP. The net result was the degradation of the Earth's ecosystems and the life-sustaining services they generate. Thus, it was necessary in the transition period to convince people that the empirical evidence was robust: beyond a certain level, more consumption does not improve our levels of well-being, and humans are not necessarily acquisitive by nature. Furthermore, it was necessary to make people understand that energy is required for all economic production and that the capacity of our atmosphere to absorb waste from fossil fuels is limited. Biophysical limits were recognized and goals changed from economic growth to qualitative improvements in human life, such as education, life expectancy, social relations and the myriad benefits generated by healthy ecosystems.

Well-being, equity, sufficiency and sustainable livelihoods were key goals that had been identified for the future global economy. They included truly sustainable consumption and production, responding to human needs and boosting the "satisfiers" to achieve them, not least food and material security. It was recognized that failed and fragile states, many embedded in turmoil and wars, cannot function without political stability, which requires economic stability and inclusive development.

The **new model of development** consistent with the vision for 2050 was based clearly on the goal of development as improvement in sustainable human well-being. This new model of development recognized that economic growth in conventional terms is a means to an end and not an end in itself. The new model focused on sufficiency of material consumption rather than ever-increasing material consumption. It used measures of progress that clearly acknowledged this broader goal. Growth was, of course, necessary in some parts of the world and for some groups, while reduction of consumption was necessary in other parts of the world and for other groups. The new paradigm acknowledged the interdependent importance of ecological sustainability, social fairness and real economic efficiency in achieving the goal of real development. It was realized that human well-being is a result of creating enlarged opportunities for people to live sustainably in both developed and developing countries, and in both economically wealthy communities and those that are financially impoverished. There was a tremendous effort over several decades to alleviate the extreme forms of poverty in every country, giving previously marginalized groups access to basic needs and opportunities to pursue sustainable livelihoods.

Ecological sustainability required recognizing that natural and social capital are not infinitely substitutable for built and human capital, and that real biophysical limits exist to the expansion of the market economy.

Social fairness required recognizing that the distribution of wealth is an important determinant of social capital and quality of life. The conventional development model, while explicitly aimed at reducing poverty, considered that the best way to do this was through growth in GDP, with a trickle-down effect. However, the growth clearly did not reduce poverty everywhere and it became clear that explicit attention to distribution issues was sorely needed. As Robert Frank (2007) argued, economic growth beyond a certain point sets up a "positional arms race" that changes the consumption context and forces everyone to consume too much of status goods (like houses and cars) at the expense of non-marketed, non-positional goods and services from natural and social capital. Increasing inequality of income actually reduces overall societal well-being, not just for the poor, but across the income spectrum.

Real economic efficiency required including all resources that affect sustainable human well-being in the allocation system, not just marketed goods and services. The market allocation system before 2010 excluded most non-marketed natural and social capital assets and services, even though they are huge contributors to human well-being. The new, sustainable ecological development model introduced as part of the pathway to achieve the vision for 2050 measured and included the contributions of natural and social capital and thus approximated real economic efficiency better.

The new development model also emphasized the correct use of market mechanisms and incentives to serve social goals, rather than letting the market lead where it will. These approaches included social entrepreneurship, fostering local markets and cooperative structures; equitable international trade regimes; re-shaping globalization; promoting appropriate technologies and sustainable production systems (in energy, sanitation, communication, food production, etc.); development of incentive/funding mechanisms for products/processes with much longer cycles; and adequately incorporating externalities in economic accounting.

The new development model also acknowledged that a complex range of property rights regimes are necessary to adequately manage the full range of resources that contribute to human well-being. For example, most natural and social capital assets are public goods. It was already known in 2010 that making them private property does not serve sustainability goals. On the other hand, leaving them as open access resources (with no property rights) does not serve those goals either. What was needed was a third way to *propertize* these resources without privatizing them. Several new (and old) common property rights systems had been proposed to achieve this goal, including various forms of common property trusts (Ostrom *et al.* 2001; Barnes 2006).

5.2.3 The knowledge and education pathway

The way that knowledge is developed or constructed and shared with others was one of the pathways by which a directed cultural shift towards sustainable development was accomplished. It was recognized that knowledge in its various forms is rarely objective, but often reflects the power, social and energetic relationships within societies.

More widespread acknowledgement of the links between knowledge, values and power in various scientific disciplines and social movements in the early twenty-first century became an opportunity for making the value basis of knowledge and education transparent. This laid the foundation for a process where there was a vigorous and widespread debate in all societies about the type of values that should underpin the construction and sharing of knowledge. The processes of generating knowledge, both scientific and other forms (in particular traditional or indigenous knowledge), and the educational programmes and activities through which that knowledge was shared (and further knowledge was generated in the very processes of education) started to give central importance to universal values underpinning care for the lives and experiences of people, other species and the ecosystems of which they are part. Other values that became influential in the construction of knowledge for technology, economy and development were equity, justice (global eco-justice) and peace.

> There are still opportunities to develop technological adaptation and mitigation strategies to face the present environmental and health challenges. Keeping pace with this, ethical values and respect for cultural and behavioural differences need to be embedded into education, science and, therefore, society.
>
> Manuel Cesário

A central strategy in this pathway became the conscious promotion of a set of values in knowledge construction and education. Among these was the valuing of diversity, in all its dimensions (ecological, cultural, etc.). This created trust; changing perceptions of what to include as "knowledge" led to new ways of generating questions and answers that affected how society acted and developed. This was a process of enhancement creating new paradigms, relational opportunities and collaborations that overcame the impasse or even apathy in attempts in earlier decades to address the world's challenges as a matter of urgency.

As formal education plays such a crucial role, much could be achieved through transforming and adapting the formal education system. Good examples of a more holistic teaching and learning approach already existed around the world in 2010. They served as models and inspiration for setting up modified or new formal education systems corresponding to the specific cultural and national context. The educational system was transformed, with a stronger focus on balancing the development of values alongside every person's intellectual, emotional, social, physical, artistic, creative and spiritual potentials. Inspiration for such approaches was found in preschool and school education such as Wood Kindergartens or outdoor youth education centres (see Boxes 5.1 and 5.2). In higher education the learning experience became more transdisciplinary, with an emphasis on collaboration and cooperation between various disciplines and stakeholders.

The reformed, or rather transformed, education system became flexible and adapted to the demands of a more connected world. This resulted in more open-minded people who valued diversity and tried to understand and respect differences between people.

> People in 2050 don't define themselves by their job titles to the extent they do today, and at the same time their lives and their work are more fully integrated. For example, Susan explained that all of her current research projects were also "courses" that she was "teaching". Teaching had become much more problem-based, "learning by doing" rather than lecture-based. Students and faculty participated in research projects together and everyone learned while attacking real problems. When I asked what department Susan was in at the university, I got that quizzical expression again. She explained that the university had become "transdisciplinary". There were research (and learning) centres focused on particular problems, but the traditional academic departments had disappeared.
>
> Robert Costanza

An equally important strategy was the informal paths of knowledge transfer. Informal education through mutual teaching and learning, achieved by gaining information through different types of media and cultural interactions, and conversation and discussion with friends or family members, plays an important role when it comes to forming people's opinions. Depending on their context, urban and rural communities developed different educational approaches that included providing information about sustainable living through community meetings and promotion days for certain topics like "sustainable and healthy food" or "saving energy".

Box 5.1 Education programmes

Howard Gardner originally identified **seven core intelligences**: *linguistic, logical-mathematical, spatial, bodily kinesthetic, musical, inter-personal and intrapersonal*; and in 1999 he added an eighth, the naturalistic intelligence. These multiple intelligences combined with knowledge cultures and systems create a multiplicity of opportunity to promote perception changes, new goals and new actions. If these new knowledge cultures were supported through educational endeavours that were reinforced across a person's lifetime, then change could occur within a generation.

Box 5.2 Wood kindergarten / Outdoor preschool

A Wood Kindergarten is a day-care for children between the ages of three and six, which was first devised in the 1950s in Denmark. Wood Kindergartens fulfil the same purpose of taking care, stimulating and educating young children as other preschool facilities. Something that distinguishes a Wood Kindergarten from other forms of preschool is that the children and day-care workers spend their time exclusively outdoors, in nature – therefore it is also known as a nature preschool. The Kindergarten is held outdoors in all seasons and under most weather conditions. For safety reasons it must provide a sheltered, heated indoor area for temperatures below –10°C, thunderstorm, wind storms, extreme snow storms or hail. Moreover, in a Wood Kindergarten, children play with toys that are objects that can be found in nature, rather than with commercial toys.

Studies have shown that this concept has a positive impact on children's development. They found that children's manual dexterity, physical coordination, tactile sensitivity and depth perception are positively influenced; e.g. such that they do not injure themselves so easily when they fall. Children who have been to a Wood Kindergarten also tend to have more highly developed creativity and communication skills as they have to verbally agree on an object function they use for their playing. Other studies have shown that staying and playing outside strengthens the immune systems of children and workers. Also, children and staff show a lower level of stress as Wood Kindergartens have lower noise levels than day-care facilities in closed rooms. All in all, a Wood Kindergarten has positive impacts on the children's development, their health and the workers' health. The environment also benefits from the concept as fewer resources in forms of housing facilities, provided toys, etc. are needed. (For more information see: www.waldkinder-ilmenau.de.)

Finally, network systems were built between universities, companies, governments, communities and individuals in order to enhance information and knowledge transfer. This enabled cooperation between specialists of various disciplines, as well as different stakeholders, users, clients, etc. Implementing all these actions created different but connected centres of knowledge accumulation than those of 2010. New knowledge cultures, which valued diversity and had balanced power relationships, were thus created.

5.2.4 The creativity pathway

The potential to be creative is something innately human, it defines us as human beings, deepens life experiences and differentiates humans from other living organisms. Creativity can take many different forms and has a tremendous potential for societal change. Here, in the concept of creativity, we include systems, institutions and informal methods that enable the creation and transmission of artistic works, entertainment, views, ideas and other forms of cultural expression. These include – among others – music, literature, journalism, art, theatre and cinema as visible manifestations. The concept is, however, more encompassing, touching also on human emotions, feelings, empathy and cultural dialogue. While considered of marginal importance in the early efforts of building a just global society, with time it became more widely recognized that tapping into the deep-seated appreciation for the creativity, passion and beauty of art as a shared language is essential for the empathy necessary to develop mutually respectful relationships between societies in a diverse world and attain self-fulfilment.

The early twenty-first century showed changing patterns of behaviour and new forms of media, which provided opportunities for greatly expanding the positive role of creative expressions for societal change. One such opportunity was the arrival and spread of the Internet, which enabled access to audiences at scales previously unheard of, and also meant that people could quite literally see how the other half lived. Online video sharing services, peer-to-peer file sharing networks and the explosion of blogging reached staggering levels, and the Internet started to change the media distribution infrastructure, as well as the relationship between individuals and communities, with new media and other forms of expression. As video and music sharing on the Internet continued to cross over from a niche market into the mainstream, there were increasing incidences of "virtual" art playing a role in the shaping of culture and societies. It gave a space for all types of traditional arts to become freely available, and allowed new forms of art to directly emerge from the virtual medium. However, even as the Internet levelled the playing field for many poor, newcomer artists, it was not until it overcame the large digital divide and became accessible for the vast majority of people – through, for example, schools and a whole chain of public Internet libraries that were institutionalized in most countries – that it truly had a global influence.

Another opportunity that was captured to support a cultural shift was the class of social icons and role models the media had created. When a few individuals in this group emerged as value-based leaders, speaking up for dire needs of people and the planet, they encouraged many to do the same, and with time the whole entertainment industry found itself as a strong protagonist of societal change, raising those difficult questions that humanity needed to ask itself.

The central strategy in the creativity pathway towards cultural change became giving creativity a much more prominent role in the education system. This was supported more widely in society through such means as state support for wider accessibility to both the means of expression (e.g. through the mass media) and access to art, encouraging people to create much more, and more diverse things; not only in the form of arts, but in science, handicrafts and lifestyle. This celebration of creativity in all its diverse forms became a tool for the intercultural communication that the world's people needed. It also reversed the trend of cultural homogenization and contributed to stronger communities that were deeply rooted in their own cultural contexts, yet greatly enriched by the wide range of creative expressions coming from all corners of the world. The focus on creativity and the associated attention this gives to inner well-being helped to redefine notions of success; moving it from material wealth to a more inclusive approach of fulfilment related to spiritual and moral values. Increasing awareness of the arts can affect the way people not only perceive art, but also how they perceive themselves, by helping to reconnect their minds to their emotions. Over time the adoption of a new appreciation of arts and emotions into the life of the community and into all sectors of society took place.

Communities started to provide new platforms of free and universal access for the display of artistic creations. At the same time, these platforms became venues for recreation such as in the form of parks and open spaces. Support systems needed for self-expression were provided by governments or other structured institutions, including funding, access to technical expertise and guidance. Partnerships and collaboration were started between all levels of the education system and the structured organizations of the artistic expression. This connection became beneficial both for education and art, bringing together potential future artists and audiences and current forms of expression. Institutional funding supported these connections and broadened their scope. By integrating art in the public spaces, shared symbols were created that strengthened the bonds of individuals and societies.

5.2.5 The values and beliefs pathway

In 2010, of all the wealthy nations, the United States had done the least to address global climate change. It also ranked last among developed nations in the percentage of its citizens (47 per cent) that believed that global climate change was caused by human activity. Only 44 per cent of Americans considered climate change to be a very serious problem, with its right-wing voters expressing less concern. Curiously, while college graduates in general were more likely to understand the causes of climate change, only 19 per cent of college-educated Republicans believed it to be influenced by humans (Pew Center 2009); thus, deeply ingrained values may be more powerful than education. In distinct contrast, the vast majority of Europeans, ranging from 78 per cent in Great Britain to 93 per cent in Spain, believed that climate change was caused by human activities, which explained Europe's growing (but still entirely inadequate) efforts to address the problem.

> They found direction and leadership from among indigenous peoples, ecological thinkers and ideas of post-democratic governance. It was recognized that the problems being experienced could not be solved by the same thinking that created them in the first place. New global democratic/biocratic governance structures were premised on the understanding that the universe is a single, integral community.
>
> Kerry Arabena

Religious organizations became strong voices for creating a shift towards an ecological well-being culture. Not only did these organizations possess authority, but they held significant financial resources, influence over billions of individuals and large groupings of communities that could act quickly in implementing changes. In parallel with the authority of formal religious organizations, there was also another major well of authority, one that typically lies untapped – that of the elders of the world. Throughout history, elders' roles have been to guide their communities, drawing on their long experience and deep wisdom gained from their long efforts in the communities and guidance from generations past. Tapping into this wisdom and energy provided a new source of power for the effort to move our societies back into being connected to the Earth and providing for all of us, not just a select few.

To balance the cultural push of counterproductive advertising, at the beginning of the transition period it became obvious that it was necessary to create an alternative cultural shift, pulling people away from increasing consumption (and defining their well-being through how much stuff they have), unless they were at the bottom of the ladder of consumption and not meeting their basic needs. Thus it was necessary to pull people towards understanding success as being responsible ecological and global citizens. This process of achieving a cultural shift was enabled through employing all kinds of measures available for societal change: spiritual and religious institutions, educational institutions, policy and governance, even counter-advertising helped to redirect our culture before the Earth redirected it for us.

A key notion underlying the directional cultural shift that was necessary to realize the long-term vision for the planet, around which there was considerable consensus in the early twenty-first century, was the broadening of the idea of wealth and prosperity. In those days, wealth had become synonymous with money in common development terms (see the discussion in Section 5.2.2). However, there was a growing recognition taking hold that societies can have wealth in a number of equally important and mutually reinforcing areas – wealth of creativity, wealth of knowledge, wealth of spiritual health, wealth of good governance, just to name a few. The shift in development came in focusing on well-being and the various dimensions which contribute to it (as discussed in the economy pathway in Section 5.2.2). In parallel there was an increasing effort to address the situation for those with low-quality lives in the high-income consumer category, who suffered from the ill effects of consumerism – obesity, lack of leisure time, long working hours and social isolation. By helping them to consume less, they had more time for the important things in life, improving their quality of life and, as an additional benefit, lowering their ecological impact and freeing up resources for improving the well-being of those who lacked sufficient material resources.

5.3 Different choices in different places

While international cooperation played a role in the transition to 2050 for dealing with problems such as climate change and ocean acidification, different countries chose different ways to achieve the vision that emphasized local and national initiatives, rather than regional or global ones.

Countries differed along a number of dimensions, which meant that different choices were made. There was no single way to achieve the vision that was best for all countries. During the transition between 2010 and 2050, countries that had traditionally focused their efforts on national and local environmental problems, or had allowed concerns for national sovereignty to impede effective international agreements and the strengthening of international institutions, had to move away from that position to one more in keeping with the global character of environmental impacts.

Related to change at the international level was the need for new institutions at other levels. While some progress had been made before 2010 in dealing with environmental problems with a comparatively limited geographic scope, many countries did not have institutions in place with sufficient support that were adequate for generating and utilizing all the required scientific and technical information. This was especially so given the context of risk, uncertainty and irreversibility within which many decisions had to be made.

The development and implementation of new and improved technologies played an important role in many countries. Some of these came from the private sector, others from the public sector, with the balance varying from country to country.

Different strategies and actions were used in different contexts to ensure a move to "sustainability" – including a convergence in consumption levels and more holistic development approaches. It took a unifying purpose and concerted strategies and actions to achieve the global, directional cultural shift which enabled us to reach as far as we have in 2050.

5.4 Responding to environmental challenges

5.4.1 Natural resources

> We need to build a new science, or at least transform the current Earth system science to an Earth-living science. If someone proposed that we should study how to live on the moon, no one would question it. But if we talk about developing a science of living on Earth, many people will view that with surprise. We should channel our Earth-living science through handbooks, manuals, textbooks, TV programmes, internet-based courses – so that those people who do not have a background in Earth system science or global change science can easily access knowledge and practice.
>
> Wenjie Dong and Peijun Shi

The major challenge for protection of natural resources was to move from facilitating univocal use of natural resources to providing guidance in dealing with (potentially conflicting) multiple uses of natural resources. Poverty alleviation and development of novel, more equitable, local options for management of natural resources was seen as a key aspect of sustainable natural resource use and avoidance of conflicts. Poor communities were often marginalized: both in terms of living in risky and remote areas subject to extraction and in terms of their political, social and economic voice. Considerable research in the social and natural sciences focused on describing such situations and the consequences for exploitation of natural resources, and much was learned from understanding and analysing historical trends concerning past successes and failures relating to policies and management of natural resources. Through embracing the insights from a wide variety of natural and social science disciplines, a common understanding of processes and trends that were driving change and conflict emerged. In addition, transdisciplinary approaches assisted through offering new options and identifying alternative resources. This formed the basis for a new approach that extended beyond the description of problems and their impact, to the analysis of future scenarios for change and the policies that would enable and facilitate equitable and sustainable use of natural resources. Energy, food, minerals and **water** are some examples of these arenas where issues of environmental governance, equity and intergenerational and intragenerational environmental justice are played out. But avoiding conflicts and human-induced disasters depended on finding a way to equitably balance local livelihoods with sustainable use.

5.4.2 Water

Some dominant regional features that can be managed to enhance freshwater resources and the self-organizing capacity of living systems are the following:

- In semi-arid regions, savannah biomes recreate the functions of freshwater in local landscapes by soil and water conservation and facilitate infiltration and plant water uptake for ecosystem services.
- In the humid tropics, rainforests safeguard freshwater functions in existing forested landscapes and maintain capacity for ecosystem services by hindering leaching of organic material and nutrients.
- In the temperate zone, the industrialized regions break the emissions of polluting substances into freshwater and groundwater and halt deterioration of aquatic food webs and ecosystems.

Malin Falkenmark

A major focus of climate change adaptation was water issues, because of the very central role that water plays in both society and ecosystems. Global warming accelerated water-related risks in large regions with dry climate, and increased the incidence of flooding in other regions, as well as in urban environments. This called for preparedness for adaptive management. The activities taken in the first half of the twenty-first century in trying to secure safe drinking water and health, water pollution avoidance and well-functioning ecosystems can be summarized as follows.

1. Safe drinking water is *basically an issue of raw water purification* and of urban water management. To reduce conflicts over water supply for

industry, demand management and waste-water reuse were introduced, especially in dry climate regions.

2 The severe water pollution, especially in fast-developing countries, called for policies to avoid downstream water supply problems, taking into account that river recovery could in fact take two generations.

3 Upstream/downstream conflicts of interest demanded basin level coordination of water uses, including both land use and ecosystem functioning in an integrated basin approach to secure water use compatibility between humans and ecosystems (Figure 5.3). Particularly in regions where climate became drier, streamflow depletion and closing river basins where all water had already been allocated introduced water shortage challenges of a new dimension, calling for research and development to find acceptable solutions. For the overarching governance, river commissions were developed. In transnational river basins, basin agreements were made sustainable so as to function under the continuous change of water use in individual nations.

4 Catchment-based land–water–ecosystem compatibility was built on the Integrated Water Resources Management approach, developed early in the transition period in countries all over the world with support from the Global Water Partnership. As Figure 5.3 shows, water input is partitioned between water for the plant cover (green water) and the surplus, forming runoff in rivers and aquifers (blue water). The concept of "green water", the water consumed by vegetation on croplands, grasslands, woodlands and forests, opened up

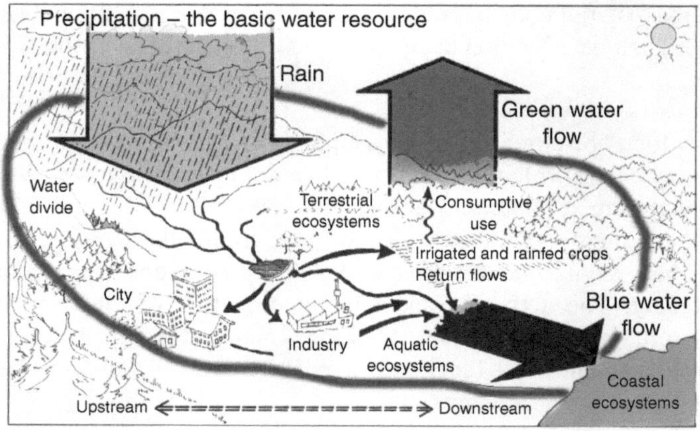

Figure 5.3 Water flow through the catchment allows an integrated approach for management (source: Falkenmark 2003).

> hydrological water-balancing approaches. These balanced land-based water use, direct societal water use and water needed by ecosystems, both the green water consumed by terrestrial ecosystems and the blue water forming habitats of aquatic ecosystems (SIWI 2005a, 2005b). In the transition period, this widened perspective of water in the landscape also pointed to the role of vapour flow in feeding the atmosphere with moisture, thus influencing precipitation patterns. Education and increasing public awareness were of central importance in the transition period, so that people became aware of the role of water in supporting all life on the planet.

Other pathways worth mentioning addressed agricultural water shortage by **water productivity increases**. In villages, cities and smallholder agriculture, water supply was expanded by **rainwater harvesting**. There was a new acceptance of **water reservoirs**, crucial to overcome the effects of rainfall variability related to increasing climate change. Increasing river pollution was abated not only by waste-water treatment, but also by dry sanitation for nutrient re-use in agriculture, so-called **productive sanitation**.

Since the water passing a river basin has so many parallel functions, pathway components towards the vision goals had to be addressed in a coordinated manner. They are all details in managing a large integrated flow pattern in which water in a basin passes in and out of human society and terrestrial and aquatic ecosystems. In this pattern, upstream/downstream conflicts of interest easily develop. **Compatibility** within the basin was secured by a coordinated approach, benefiting from the integrating capability of water moving towards the river mouth. The goal was **eco-hydrosolidarity** between all human activities and all water-dependent ecosystems, terrestrial as well as aquatic. A key role in these efforts was played by **River Basin Committees** for joint decision-making in the basins, supported by **stakeholder participation** to ensure that decisions were socially acceptable.

5.4.3 Agriculture and land use

Achieving food security involved major challenges, in view of the congruence between under-nutrition, dry hydroclimate, poverty and continuing population growth. Many countries with large populations depend on food imports for adequate diet levels, a fact that influences national economic planning in poor countries without purchasing power. In some of these countries, cropland expansion was also necessary. Overall, for the agriculture sector, there was a move towards a land-system view, away from a production-system view in order to achieve a more integrated approach and thus achieve the vision for 2050.

It was possible to meet the challenge of feeding nine billion people sustainably in 2050 partly through changes in the agricultural system. First it was necessary to channel capital and intellectual investment into agriculture in ways that ensured equitable outcomes for food production, ecosystems balance and global sustainability. This required knowledge about living beings, environments, technologies and socio-economic aspects, as well as the integration of different scientific disciplines and the multiple objectives of agriculture, actors, space and time-scales, to design and implement sustainable food and agricultural systems towards 2050. The pathway towards these agricultural and food system ideals was achieved by:

- reducing waste and losses at all levels: field, storage, transport, processing, distribution, consumption;
- balancing food consumption trends and sustainability conflicts, especially those relating to the association between higher income and increased consumption and vice versa through responsible consumption behaviour;
- developing better ways of utilizing urban-rural landscapes for food systems and natural resource use;
- investing in research and development and training in order to increase yields efficiently;
- designing and distributing agricultural production systems compatible with the social and ecosystems goals in ways that contribute to preempting and managing climate change and its impacts;
- making appropriate social, technological and economic investments in infrastructure and regulation of agricultural trade, in ways that promoted redress of vulnerabilities and cushioned areas and communities that continued to suffer food shortages.

At the beginning of the transition period it was possible to apply existing knowledge and technology, as well as integrated agricultural research for development. Appropriate social and environmental food labelling and certification of agricultural foods and services, as well as awareness and education programmes, enabled responsible production and consumption. The core components of a sustainable agricultural growth strategy were technological change that increased output and lowered production costs; investment in rural infrastructure (such as roads) that reduced transportation costs; widespread participation in agricultural markets by smallholding peasant producers and women farmers; and significant reforms in natural resource management. After the transition was underway, implementation of knowledge, solutions and research findings to bolster new perspectives on food production were achieved through an integrated approach to knowledge and technology management, capacity building, making science relevant to society and instilling appropriate value systems. Other developments that drove the realization of this subtle agriculture and food system entailed an upgrading of irrigated land through appropriate technology and a recreation of institutional frameworks to support it, including the establishment of global regulatory systems/bodies/institutions with the mandate to oversee the elements of this pathway.

5.4.4 Bioenergy

It was recognized at the start of the transition period that bioenergy needed to contribute positively to offsetting fossil fuels and to rural livelihoods (Hall and Overend 1987). At the same time, however, it was realized that a rapid acceleration of biomass production for energy use could exacerbate degradation of land and ecosystems (Searchinger *et al.* 2008), increase GHG emissions and increase food prices and subsequently reduce food security. Thus, in order for biofuel plantations to contribute to sustainable development, agriculturally sustainable methods were required and markets that provide enhanced livelihoods had to be developed. This was not happening before the transition began. On the contrary socio-environmental effects were associated with sugarcane ethanol systems in Brazil (Martinelli and Filoso 2008) and serious conflicts between the production of biofuels and food production emerged in the years before 2010 (Sagar and Kartha 2007). The challenge thus lay in translating the opportunity into reality.

In order to address this challenge, all negative consequences of biofuel production had to be better understood and communicated. In addition, an appropriate regulatory structure had to be found. Only then could decision-makers set more dependable targets and develop effective legislation to deny unsustainably produced biomass access to the markets. Product certification that guaranteed production compliance with sustainability goals, for example, contributed to this (Zarrilli and Burnett 2008) and both producers and consumers accepted the value of these schemes. Research also rapidly advanced the development and use of next-generation bioenergy, which was independent of food crops and less dependent on land. Fermentation of cellulose that is amply available in wood and agricultural waste (Himmel *et al.* 2007) or production of oil from fast and continuously growing algae (Gao and McKinley 1994; Lewis and Nocera 2006) were among the approaches. They became available at industrial scales within a few decades, and synergies with, for example, power-plants with excess heat, CO_2 and in some cases also necessary nutrients could easily be obtained. These advanced technologies could also be used locally to treat existing organic waste streams and thus helped in improving sanitation in urban areas. These advanced, integrated bioenergy–food systems had to incorporate landscape and watershed patterns of resource usage, recycling of materials back to ecosystems to maintain nutrient and water resources, and protection of ecosystem services and other aspects of biodiversity. Development of biorefining technologies also had to consider vertical and lateral integration of ecosystem services and landscape interconnections.

5.4.5 *Reducing emissions of carbon dioxide*

> Technology research and development will be important for altering the emission trends. Solving global environmental problems will require global diffusion and transfer of new as well as existing technologies, recognizing "common but differentiated responsibilities".
>
> Washington O. Ochola

Emission reductions were achieved through a carbon tax increasing over time – revenue-neutral to subsidize travel costs of low-income workers – and applied at mines, wellheads and ports of entry, combined with a moratorium on new conventional coal plants (Hansen 2008). Furthermore, it was agreed that the most effective strategy was direct investment by governments and the private sector in needed energy research, development, demonstration and deployment, as well as a major focus on improving energy efficiency, for which significant institutional issues had to be resolved. This strategy was in effect an Apollo-like programme that provided "green jobs" in twenty-first-century industries (Hoffert *et al.* 1998; Gore 2008; Shellenberger and Nordhaus 2009).

5.4.6 *The transitions of urban areas*

Urban areas were the subject of a large number of studies covering their diverse dimensions following disciplinary perspectives (urban economy, urban politics, urban sociology, urban anthropology, urban geography, urban climate, urban hydrology, urban ecology, etc.). Other valuable and available domains of knowledge on urban areas include local knowledge (urban and architectural design, the local landscape) and the knowledge of local urban practitioners and policy-makers. All these domains of available knowledge created a broad base to support **new approaches to urban areas**. A major challenge was to improve urban water supply for the more than two billion inhabitants living in cities in dry areas.

The pathway of knowledge and education (see Section 5.2.3) was thus very important to reaching the 2050 vision. Creating new integrated and multidimensional knowledge of urban areas was useful for identifying and avoiding the negative consequences of their bidirectional interactions with global change. These integrated perspectives of urban areas also helped to define new paths of urban growth, and restructuring and improving institutions addressing urban growth. An effort to engage local practitioners (urban planners and policy-makers) by international and national organizations offered the opportunity to provide a large number of practitioners with multidimensional knowledge of the connections between local urban and environmental problems with global and regional environmental processes. Increasing wealth facilitated urban governments' abilities to provide education, increasing the capability to develop technologies and help with the costs of knowledge integration.

Cities provided enormous opportunities for all pathways due to the proximity of human beings for the diffusion of innovation, creativity and values. While slums and poverty created constraints for following various pathways to sustainable development, they also provided opportunities. Slums provided enhanced environments for the development of informal economies and for formal economic activities that enhanced quality of life. Low land-values opened up opportunities for the use of land for valuable publicly supplied support systems (such as new biofilters that remove much of the nutrient content from urinal flows in a vertical system). Slums enabled the social capital that enhanced resilience to economic and other shocks. Finally, slums were often transformed into well-organized urban spaces over time and incorporated into the formal urban area. They were the basis for the creation of the urban space in many poor countries and emerging economies, and even in some urban areas in rich countries. This is important when one considers that most urbanization during the transition period occurred in emerging economies and poor countries.

Urban areas are centres of creativity and innovation, which meant that the creativity pathway (see Section 5.2.4) was very important in these areas. In 2010 many cities already offered examples of technological innovation and other creative initiatives to reduce pressure on the environment and enhance human well-being. Examples include: Curitiba, Brazil, which since the 1990s had aimed to create a more balanced relationship between the urban area and the environment, as well as an improved quality of life for its inhabitants; European cities that had introduced car-sharing systems; and green roofs in Malmö that were introduced to deal with storm water. It was important that information on experience with innovations could be exchanged freely and in a way that encouraged other urban areas to experiment with similar initiatives. Education played a central role in increasing the capacity for innovation and creativity, as well as the ability to assess the risks of new technologies in a systematic way. While institutional frameworks that supported innovation and diffusion of eco-efficient technologies played a significant role, everyone living and working in the urban environment had to reflect on their real needs for ecosystem services and the best ways for satisfying those needs.

While many urban areas already had a public transportation network, at the beginning of the transition period the increasing personal use of cars continued to add to emissions of GHGs, building of roads and car parks and air and noise pollution. Improving public transportation in urban areas reduced environmental pressures and had substantial economic and social benefits. This improvement made use of existing knowledge, for example, on how to provide incentives for the public to switch to public transportation. Education and outreach played an essential role in encouraging the development of public transportation, for instance by enhancing awareness of the multiple environmental and health benefits of reducing emissions. At the same time, research and development was required for the creation of efficient technologies to be used in public transportation. Value changes, guided by realistic fuel pricing and congestion charges, and supported by equitable charges for the provision of widespread and easily accessible public transportation, meant that the public could make long-term decisions on their preferred choice for mobility services, including "low-technology" solutions and decisions to reduce mobility.

The growing concentration of population and the economy in urban areas during the period between 2010 and 2050 opened up the opportunity to change consumption patterns (food, building materials, consumer goods) with an impact on global, regional and local land use. With the knowledge pathway, studies of urban metabolism and the ecological footprint approaches provided useful information about the fluxes of energy, resources and materials to urban areas. With the pathway of education, the concentration of population in urban areas had a multiplying effect in creating new consumer cultures and awareness of the consequences of patterns of consumption. More people could be reached more easily through education and outreach programmes (public authorities, professional organizations, NGOs, religious groups, private sector, media) in urban areas. Furthermore, the use of available environmental technologies reduced the demand for natural resources and environmental pressures, and improved human well-being in urban areas. This technology was easily disseminated among different social groups in urban areas.

5.5 Conclusion

The challenge that the world faced in 2010 was immense. A number of pathways were needed in order to achieve a much more sustainable situation in 2050. It was important to recognize that we were dealing with an extremely complex system in which there was much uncertainty, stakes were high, decisions were urgent and values were contested.

Our world in 2050 is indeed a much better one than the world of humanity's past. It has gone a long way towards many of the long-term goals around which humanity has stood united, but which took deep crises and considerable concerted efforts to move towards. These efforts were concentrated along five major pathways, in each of which fundamental transitions made way for development: governance and institutions centred around core values and principles; reshaping of the building blocks of the global economy; a vastly expanded education and knowledge-generation system; tapping into the enormous source of human creativity; and harnessing the inspiration for action in value and belief systems of the world. Each of these pathways represents areas in which structures, institutions and informal collaborations have supported progress towards the better world of 2050, and at the same time vitally connected human and societal health to ecosystem health.

> These strategies will need to deal with a multi-sector and multi-scaled set of issues simultaneously. These include energy, thresholds and abrupt changes, development strategies, resource-sharing and risk assessment.
>
> Dennis Ojima

We started from many different places and faced many different choices on how best to proceed. There was no single pathway towards humanity's long-term goals that fitted all societies, nor was any one pathway linear. Progress was made by different societies in different ways, with each society making choices that best reflected their own priorities and that addressed their strengths and weaknesses directly, while remaining aware of their global responsibility. These choices, and the resulting progress towards the common long-term goals became the core of "development" for humanity.

This notion of development as it came into being was dependent on fundamentally redefining the relationship between the prevailing concept of development in 2010 and the global economy. During the twentieth and early twenty-first centuries, economic growth was largely viewed as the key pathway to reducing poverty, improving health, increasing education and, in general, enabling better lives. It became increasingly obvious that economic growth did not necessarily lead to better, richer lives. After the economic depression that started in 2009, it was realized that in order to really achieve human development, modern notions of progress had to be decoupled from monetary wealth generation. The global economy was repositioned in the global development landscape. Instead of having primacy in the framework and being seen as the basic driver towards other improvements in society, economic development came to be seen for what it is – only one of the mechanisms for creating better lives.

The deep transitions taking place in the economic system and in the other four pathways of governance, creativity, knowledge and education and values and belief systems were enabled by and contributed to a directional worldwide cultural shift; that is, a shift in values, norms, habits and structures of society. One important shift in values was a recognition of the central role that water plays in maintaining all life on the planet.

We have briefly explored technological, institutional and conceptual approaches to solving the issues of climate change and peak oil. Again, this reflected the recognition that no single approach would suffice. Technological change was essential. It was clear that it would not be possible to prevent climate catastrophe and simultaneously meet basic human needs with the fossil fuel technologies existing in 2010. However, without changes in paradigms and goals, it would have been impossible to muster the political will necessary to promote the institutional and policy changes required for large-scale implementation of existing technologies and essential research and development into new ones. Worldviews, institutions and technologies had to work together synergistically, and it was necessary to rely on changes in all three to create the new energy system of 2050.

It was clearly necessary to adopt an adaptive management approach to achieve our vision (Norton 2005). It was never known for sure how well or how fast any policy, institution or technology would work, or what was the best way to change world views. Given the scale and urgency of the problem, it was important to experiment with a variety of approaches, testing our hypotheses concerning their effectiveness. Each experiment gave us more knowledge, allowing us to develop better pathways in an iterative process.

6 Back to the reality of 2010 and the challenges ahead

6.1 Introduction

As we have seen in Chapter 3, the state of the planet in 2010 is far from satisfactory. Not only is there ample evidence of a deteriorating environment, we also see weaknesses of governance, problems with the global economy and a spreading culture of consumerism that means that globally we use more natural resources per year than the planet can sustainably produce. In the recent past, society has responded through partial solutions to specific problems without a systematic and careful consideration of impacts, or effectiveness with respect to an integrative approach to the inherent coupling between human and environmental systems.

> Phase-out of fossil fuel CO_2 emissions by mid-century as global energy demand doubles or triples is the engineering challenge of the century; but doable in principle.
>
> Marty Hoffert

The vision for 2050 is truly ambitious, but by outlining a world that is quite different from that of 2010, we are setting a goal that provides a focus for dialogue in society. The pathways to the vision described in Chapter 5 are equally ambitious and demonstrate the kinds of deep structural change that will be necessary to achieve the vision. As we described those pathways, we noted a number of challenges that will have to be tackled. This chapter summarizes those challenges.

Meeting the challenges and dealing with them in a way that is effective and fits with our overall vision of an equitable and sustainable future will require **strong leadership** supported by, and even pushed by, a society that is fully aware of the need for change and willing to accept some of the inevitable trade-offs.

> No one part of society can tackle these issues on its own. We know this for sure. We also know that smart, integrated solutions are indispensable, bottom-up as well as top-down, in developing as well as developed countries. Of course this requires clarity of purpose, great leadership, understanding of roles and responsibilities, freedom and motivation to act. It also requires compelling inspiration to engage one way or the other as individuals and communities, to do one's bit, or fulfil obligations, as part of a whole and for future generations, rather than just exercising rights. Some call this moral purpose, others the equivalent of war-time mobilization.
>
> Anthony Simon

Achieving the vision along the pathways described in this book will also challenge the scientific community itself. Placing a stronger emphasis on involving stakeholders in the formulation of research questions and development of iterative, goal-searching processes that are strongly implementation-oriented will require major changes in how science is organized, funded and evaluated.

6.2 The challenge of education and awareness-building

In 2010 we have education systems that are often bureaucratically organized. The routines and mindsets of educators and their high level of specialization are reasons for resistance to change in the organization of education. The construction of formal education systems defines not only knowledge, but also what is considered to be "ignorance". The result is narrowly defined knowledge and specialization, which leads to imbalance in valuing specific types of knowledge. This specialized knowledge is socially accepted knowledge and therefore trickles down to mutual teaching and learning. A related challenge is that holistic knowledge, as opposed to specialized knowledge, is perceived as extreme and radical. Advocates of holistic thinking often criticize the prevailing ways of living – especially in the western world – and call for a new mentality.

Another challenge is the fact that education – especially in industrialized western societies – has focused on creation of skills and professions specifically within fields that advance technology, economy and development. This has occurred at the expense of development of moral and aesthetic senses that are fundamental to ideas of equity, justice and peace in the world. The education systems thus do not promote and help build the elements of well-being, strengthen character development and improve the capacity for societal relationships through, for example, the tapping of creativity in various artistic forms.

A reform of the education system is therefore a prerequisite for the transformations required to achieve the vision of 2050. Since the adults of 2050 are being educated from now on, we also need to emphasize life-long learning.

If we are lucky, very lucky, political will for clean energy technology transfer will be transparent, with the successful deployment of clean energy technologies from entrepreneurial and developed nations to those in need. This can only be attained through education, education and education. As long as the constituency is ill-informed, irrational, and impatient with complex issues, poorly formed decisions will be the outcome of any future.

Kathy Hibbard

6.3 The challenges of creativity

In the twentieth and early twenty-first century, art, creativity and emotions have been pushed farther and farther away from the core of societal structures. Institutions and norms, including the education system, as discussed above, support the preservation of the hierarchy of overly materialistic priorities in society. This represents an important obstacle to using creative expressions as a pathway towards the fundamental cultural and societal changes that are needed.

Another challenge for harnessing creativity for cultural change is the lack of a common understanding or appreciation of art in society. Each person sees expressions of creativity through the filter of her or his own experiences, emotions and expectations, which means that the influence of creative expressions is not uniform. What for some is art or free expression, for others means nothing or can even be felt as an insult. Ignoring this fact has led to an atmosphere of mistrust between different groups of society, which many times impedes the true building of a meaningful cultural dialogue and thereby cultural change.

Pop culture has been a dominating influence on many generations. It was a product of an industry in its own right, designed not to educate but to entertain. It has had a very important role in shaping the characters and personalities primarily of young people, providing both behavioural models and role models. But as it was produced to be sold and to be consumed by as many people as possible, it was designed to address the most common and basic psychological needs, and reduced people only to their role as consumers. Moreover, it has also led to homogenization of individuals within and between societies and a decreasing role of other, non-western cultural expressions.

Harnessing creativity for cultural change is a major challenge for the achievement of the ambitious vision for 2050. Meeting this challenge will require serious engagement with monolithic institutions, including the advertising industry, as well as societal dialogue about prevailing norms. Moreover, since the true challenge of creativity is borne by each individual in their life and their view of the world, it will be important to enhance the possibilities for people to unleash their creativity at various stages during their life and learn to be creative in solving problems.

> Importantly, however, radical changes in social organization are needed to realize this desirable vision. Not only will these changes be difficult and complex, they will necessarily be different from region to region and locality to locality around the globe. While efficient technologies may be more universal in application (although this is also a debatable claim) socio-economic and socio-ecological structures are not. What will therefore work in New York City, may not work in Shanghai and what works in Durban may not work in Panama City.
>
> Peter Marcotullio

6.4 The challenge of the global economy

At present in the industrialized world there is a prevailing faith in the ability of the market to solve resource problems. Market prices are believed to balance supply (costs) with demand (benefits). Government policies therefore tend to default to "market" mechanisms of penalties, payments and property rights.

> Labelling climate change and other problems resulting from the interaction between social and biophysical processes only as environmental problems hinders the creation of structural approaches to address them. They are also development problems. Societies are used to addressing environmental problems through technical solutions on an ad hoc basis, seldom focusing on their structural causes and the broad range of their consequences. They receive also a lower priority than economic growth.
>
> Roberto Sanchez-Rodriguez

From a markets perspective, global climate change is viewed as the result of correctable market failures. No one currently owns the waste absorption capacity for CO_2, which results in over-use by everyone – the inappropriately labelled "tragedy of the commons" (Hardin 1968), which is really the tragedy of open access (see also Chapter 4). To let the market work its magic, governments can create tradable property rights (a cap-and-trade system). Another diagnosis is that climate change is the result of negative externalities of fossil fuel use. In this case, carbon taxes force economic actors to internalize these external costs into their decisions, leading to socially optimal emission levels. Yet a third diagnosis is that global climate stability is a public good, which markets fail to provide. In this case, subsidies for developing and adopting technologies that reduce carbon emissions are appropriate. Unfortunately, most existing government subsidies perversely encourage fossil fuel use. With payments and penalties, prices determine emission levels, while with cap-and-trade, allowed emissions levels determine prices (Daly 1997). National governments, primarily European, are experimenting with cap-and-trade, green taxes, feed-in tariffs (a guaranteed higher-than-market price for alternative energy, i.e. payments) and so on. The Kyoto Protocol is an intergovernmental global institution seeking to implement cap-and-trade.

Payments, penalties, caps and market prices are leverage points that Meadows (1999) lumps together as "numbers". Unless set high enough to create feedback loops appropriate to the scale of the problem, numbers are the weakest of Meadow's levers for intervention in a complex system, explaining the failure of prices to solve the peak oil problem and of penalties and payments to have any effect on climate change.

However, even if we could muster the political will to set these numbers at appropriate levels, **markets are entirely inadequate to realize our vision**. In addition to measuring scarcity, the price mechanism in theory has two other major functions: it allocates resources towards those products that can be sold for the greatest profit, thus maximizing monetary value on the producer side; and it rations products towards those individuals willing to pay the most, thus maximizing monetary value on the consumer side. Unfortunately, markets often fail to maximize monetary value, and even when they do, maximization of monetary value will fail to create the energy future we envision.

Markets fail to maximize market value in several ways. Perhaps most obvious, there are numerous well-documented circumstances where firms and households fail to adopt existing energy-efficient technologies, even when annual returns on investment are as much as 100 per cent. Price incentives are highly imperfect. Furthermore, the rationing and allocation functions fail to function correctly with information, the source of the new technologies we desperately need. When the creation of new technologies is left up to market forces, they are patented then rationed by the price system to those who can afford to pay. Information, however, improves through use, so price rationing creates artificial scarcity (Daly and Farley 2003). Concrete examples are given in Box 6.1.

Markets may also be an inefficient way to generate new information. Market research and development is competitive, with each corporation striving to be the first to develop and patent a new technology. Scientists involved in developing new energy technologies regularly point out the importance of the global scientific community in cooperatively developing new technologies. New breakthrough technologies rarely appear in a short time-scale from market-driven systems.

Even when markets do maximize monetary value, this does little to provide adequate energy for the world's poor to lead comfortable and productive lives. Monetary value is determined by market demand, which is calculated via preferences weighted by income. Markets provide little incentive to provide energy (or other benefits) for the poor (see Box 6.1 for concrete examples).

Box 6.1 Markets as an obstacle

Markets as an obstacle to the advance of knowledge

The Convention on Biodiversity awards property rights to genetic information to countries in which that information is found. When a new strain of the avian influenza recently appeared in Indonesia, the Indonesian government was concerned that if a private corporation developed and patented a cure, it would be unaffordable to the Indonesians. Indonesia therefore sold a strain of the flu virus to a private corporation in exchange for access to any resulting vaccine, which prevented other firms from using the genetic information in the strain. One firm has less chance of finding a cure than many firms would have. If a cure is found, it will be patented, and prices will ration use to those who can afford it, creating artificial scarcity and increasing the chance of a global pandemic.

Markets as an obstacle to the dissemination of knowledge

Patents on non-ozone-depleting compounds raise their price, contributing to increasing use of ozone-depleting HCFCs in China and India, and a record-breaking ozone hole in 2006 (Farley 2008). In an analogous fashion, if market forces develop inexpensive, safe and carbon-neutral energy technologies, patents will increase their prices, which paradoxically decreases their value to society (though not to their creators, of course). If poor countries cannot afford the technologies, they will continue to burn coal.

Market as an obstacle to meeting the needs of the poor

Aventis developed a compound, eflornithine, that killed trypanosomes, the parasite responsible for African sleeping sickness. Aventis could not profit from selling the drug to poor Africans and discontinued production, but licensed the drug to Bristol Myers Squibb (BMS) and Gillette to remove unwanted facial hair in women, a very profitable enterprise. Although Aventis and BMS agreed to again produce eflornithine for the treatment of African sleeping sickness after the NGO Médecins Sans Frontières threatened to publicize the issue, had the market been left to its own devices, the rationing function of price would have continued apportioning eflornithine to hairy women rather than destitute Africans. Just as markets allocate far more resources towards industries that supply cosmetics for the rich rather than those that supply cures for lethal and contagious diseases that afflict the poor, they similarly allocate resources towards developing energy technologies for the luxuries of the rich over the necessities of the poor.

Clearly, a reliance on markets as a solution to resource (and other) problems will not support the development of the vision. Innovative solutions are needed, such as those suggested in Chapter 5. These innovative solutions also address other major issues related to sustainability, such as employment and work–life balance.

6.5 The challenge of time-scale

Even if we could eliminate all human perturbation of the Earth system today, it would simply not be possible for various functions and processes within the system to come within desirable limits by the year 2050. Many of the processes of the natural world operate on long time-scales. Emissions of sulphur (the major cause of the acute acid rain problem of the twentieth century) were cut radically across Europe and North America through the 1980s in response to health and environmental concerns, but in many regions, the recovery of the aquatic and land ecosystems only became evident two or three decades later. The circulation of the ocean, a complex process shaped by regional differences in water temperature and salinity, is a centuries-long process too. This has implications for the re-equilibration of the anthropogenic carbon dioxide between the atmosphere, land and ocean: past emissions from fossil fuel burning and land-use change have already committed humanity to significant temperature change (e.g. Ramanathan and Feng 2008).

The slow processes of recovery mean that the improvements of the environment will not be complete by 2050, even if major efforts are started today. This could create misconceptions and indeed empower sceptics looking for more immediate results for the "sacrifices" they are asked to make. So, increasing public awareness about the slow speed of recovery as a result of complex natural processes will be necessary.

6.6 The challenge of complexity and irreversibility

Some changes that human activities have induced are irreversible, such as losses in biodiversity – the destruction of populations, entire species and habitats. Therefore, it must be recognized that it will never be possible to restore all parts of the Earth system to former states.

Furthermore, both science and practice have demonstrated that the closely interconnected human and nature systems do not follow linear, predictable and controllable paths. In particular, it is clear that human choices that we cannot predict have a bearing on the pathways that will be taken. We now recognize that socio-ecological systems exhibit a high degree of non-linear behaviour. This is true for all types of systems, from local ecosystems, to regional systems and global systems.

Complexity can also be a challenge to dealing with issues that are part of our vision for 2050. For example, control of emissions of non-CO_2 GHGs (i.e. methane, ozone, nitrous oxide) is exceedingly difficult for a number of reasons but, most importantly, because many of these emissions are related to global food production/security. Therefore, in order to keep global warming in check but at the same time improve food security, emphasis must first be placed on reducing CO_2 emissions from fossil fuels to zero by 2050 by transforming the world's energy systems so that renewable sources of energy and energy conservation play a dominant role.

> Competing claims on natural resources become increasingly acute, with the poor being most vulnerable to continuing global change. The major challenge is to move from facilitating univocal use of natural resources to providing guidance in dealing with (potentially conflicting) multiple-uses of natural resources. Poverty alleviation and development of novel, more equitable, local options for management of natural resources is seen as a key aspect of sustainable natural resource use and avoidance of conflicts.
>
> Mary Scholes

6.7 Institutional challenges (with an emphasis on the energy sector)

Leading systems thinker Donella Meadows identified several powerful levers for effecting change in complex systems (Meadows 1999). Four powerful levers are: the rules that guide a system, information flows, feedback loops and numbers. Another term for "rules that guide a system" is **institutions**. Institutions play a significant role in all of the pathways discussed in Chapter 5.

Changes in institutions and policies are essential to achieve the vision for energy in 2050. Because of the particular physical characteristics of the resources involved, the solutions to the peak oil and climate change crises require cooperation, not competition. The critical ingredient in developing new clean energy sources, however, is information. Rather than private corporations competing to produce privately owned information, the energy challenges will have to be met by cooperatively produced, open-access energy technologies. Ironically, while the open access nature of planetary waste absorption capacity for CO_2 is one major cause of global climate change, private property rights to clean energy technologies are a major obstacle to addressing it – a "tragedy of the non-commons" (Farley, in press).

> Two approaches can be identified. The first is a material approach, using science and technology to reduce the negative impact on the environment from human activity. The focus is on developing technology and social organizations that require fewer raw materials and pollute less. The second approach is on the spirit, attempting to reduce the appetite and carelessness which drives much of the damage we inflict on the environment. This is also essential, but given the material hardship and desperation among some 60 per cent of the world's population that drives much of the damage to the environment, it will be extremely difficult. A combined positive–negative approach will be necessary.
>
> Tomas Ries

Energy is such an important resource that the dominant source of energy can have a major influence on the economic system. The transition to a clean and safe energy future, well supplied with ecosystem services (which for the most part are also public goods), requires that substantial portions of the economic system be based on cooperation and common ownership rather than competition and private ownership.

This change of ownership patterns could be achieved through the creation of a global common asset trust (GCAT) with explicit property rights to the shared inheritance of humankind, and an explicit mandate implemented by a board of trustees to manage it for this and future generations. The mandate would carry the same weight as a political constitution, helping to insulate the trustees from corporate influence. The GCAT would require a global agreement made by existing governments.

Further policy tools will be needed for the transition of the energy system. These include a moratorium on new conventional coal plants (Hansen 2008) and efficiency mandates. Carbon taxes could be set to increase over time, and would have to be large enough to create powerful feedback loops. One possibility is to cap use and rent out usufruct rights for short time periods to the highest bidder, which would mean that caps had to be determined by ecological limits, not short-term political expediency. Revenue from taxes or caps would be recycled back into massive investments in needed energy research, development, demonstration and deployment, creating a "new green economy" (Hoffert *et al.* 1998; Gore 2008; Shellenberger and Nordhaus 2009). To be most effective, the resulting technologies should be open access. Once appropriate technologies have been developed, market firms could compete to produce these technologies as cheaply as possible. Effective policies would stimulate the market to do what it does best, while clearly acknowledging its shortcomings.

The plausibility of global cooperation has begun to increase at the beginning of the transition period. The financial crisis of 2008/09 has led to calls for a "new Bretton Woods" and calls for a "Green New Deal". The original international Bretton Woods agreement made in 1944 sought to facilitate the global production of market goods, to prevent economic depressions and reduce the risk of world war. The new Bretton Woods would instead facilitate the provision of global public goods (such as climate stability and other ecosystem services) through cooperative efforts. This effort would aim to reduce the risk of war and hostility by reducing resource conflicts and by stimulating trust, bonding and more cooperation. The GCAT could emerge from the new Bretton Woods as a global funding source for critical technologies, and any government or institution could start such an effort unilaterally, with open doors for any other nations or institutions that care to join.

Finally, the energy and climate change crises demand coordination across scales, from the local to the global. Appropriate political and economic institutions must be based on the principle of subsidiarity, with decisions made at the most decentralized level possible, but no lower. Climate change is obviously a global problem, and mitigation demands stronger global institutions than exist at the start of the transition period – although it certainly is possible to build on existing institutions, such as the Kyoto Protocol. Global institutions set macro-level goals, such as targets for global emissions. How local communities meet those macro-level goals must, however, be flexible for two reasons. First, while studies suggest that basic human needs are constant, how we satisfy them differs (Max-Neef 1991), and local institutions are best tailored to deciding precisely what steps local communities should take to achieving global goals. Second, no matter how quickly we act, some climate change is unavoidable. Adaptation needs are very different in different locations, and while it may be facilitated by global cooperation, many decisions have to be made at the local level. We need micro-level flexibility in order to achieve macro-level goals.

6.8 The overriding challenge

Our challenge is to engage the very cognitive processes that have got us where we are today. Some cultures over centuries have internalized a belief that people should have dominion over nature (Cronon 1996). The necessary rebalancing of the interaction of our own species with the rest of nature will depend on a reversal of a very widespread and long-standing detachment of humans from nature.

> We don't understand the socio-ecological system very well – not as an interdependent and dynamic entity, although there are communities that understand its two components fairly well. We can't predict it. We have tended to avoid discussing where that kind of activity would lead us – too metaphysical, too political. Too difficult, given the language and culture barriers between the separate expert communities. In holding onto the idea that science might be "value free", most scientists are out of the habit of reflecting on whether their knowledge is good, or the uses it's put to are right. (Of course, there are some who are expert and manipulative players of the political game.)
>
> Sarah Cornell

We need to develop memory, not just on the emotional time-scale of a human lifetime, but we must learn to "think back" over our species' lifetime, through historic records, archaeology and anthropology. Furthermore, using emerging scientific methods for palaeoclimate reconstruction and geology, we must "remember" further back, to include rare, gradual and recurring climatic events.

> The environmental crisis is very much also a question of justice – intergenerational and intragenerational justice and justice towards

nature. In his famous "Free at Last" speech in 1963, Martin Luther King said, "We have also come to this hallowed spot to remind the world of the fierce urgency of now. This is no time to engage in the luxury of cooling off or to take the tranquilizing drug of gradualism." His dream for civil rights has come through almost entirely and this should be inspirational for us in the environmental movement as well.

Sofia Guedes Vaz

Furthermore, we should not forget that the major transitions between now and our vision for 2050 will bring new challenges not foreseen at the present time, so we should prepare for those as well. And finally, it is important to bear in mind that the journey does not end in 2050: the challenges in 2050 remain tremendous.

The task is immense and will require multi-disciplinary approaches. The tasks must involve the international community with participation, not only of the richest countries, but also of the emerging world, which will play a key role in the future solutions. The approach will have to be global, but with the recognition that changes and their impacts occur primarily at the regional and even local levels. The dialogue that exists between the community of scientists and of policymakers needs to be considerably enhanced, and the appropriate tools to translate the results of the research into accessible information for society need to be developed. Just like 50 years ago when the scientific community decided to develop societally important weather prediction models, it is imperative today to establish an observational and modelling infrastructure that will monitor and predict the evolution of "patient Earth" and assess its likely response to possible actions by society. The "medication" ranges from changes in the behaviour of the population, the adoption of new policies, to more controversial geo-engineering approaches.

Guy Brasseur

Notes

2 Developing a vision for the planet

1 They were reprinted in "Deeper News", published by the Global Business Network (www.gbn.org), with an introduction by Adam Kahane, who facilitated the scenario process.
2 www.australia2020.gov.au.
3 www.uvm.edu/giee/ESDA.
4 www.nif.org.
5 www.countryside.gov.uk/LAR/archive/Quality/overview/index.asp.
6 www.wbcsd.org/Plugins/DocSearch/details.asp?DocTypeId=25&ObjectId=MzczOTc.
7 www.worldfuturecouncil.org/a_renewable_world.html.
8 www.europeanclimate.org.

3 The starting point: planet 2010

1 www.un.org/documents/ga/conf151/aconf15126–1annex1.htm.
2 www.un.org/documents/ga/res/41/a41r128.htm.
3 www.un.org/womenwatch/daw/beijing/platform/declar.htm.
4 www.un.org/millenniumgoals.
5 www.arcworld.org/news_headlines.asp.

5 Pathways between 2010 and 2050

1 Strategy Unit, Cabinet Office, Government of the United Kingdom (2008) *Achieving Cultural Change: A Framework*. www.cabinetoffice.gov.uk/media/cabinetoffice/strategy/assets/achieving_culture_change.pdf

References

Amin, S. and Graham, S. (1999) "Cities of connection and disconnection". In: *Unsettling Cities.* Allen, J., Massey, D. and Pryke, M. (eds). London: Routledge.

Backlund, P., Janetos, A. and Schimel, D. (2008) *The Effects of Climate Change on Agriculture, Land Resources, Water Resources, and Biodiversity in the United States*. Report by the US Climate Change Science Program and the Subcommittee on Global Change Research. purl.access.gpo.gov/GPO/LPS95356 (accessed 1 March 2010).

Barnes, P. (2006) "From common wealth to common property". www.vtcommons.org/journal/2006/01/common-wealth-common-property-peter-barnes.

Bernstein, L., Roy, J., Delhotal, K.C., Harnisch, J., Matsuhashi, R., Price, L., Tanaka, K., Worrell, E., Yamba, F. and Fengqi, Z. (2007) Industry. In: *Climate Change 2007: Mitigation.* Contribution of Working Group III to the Fourth Assessment Report of the Intergovernmental Panel on Climate Change. Metz, B., Davidson, O.R., Bosch, P.R., Dave, R. and Meyer, L.A. (eds). Cambridge: Cambridge University Press.

Bradshaw, Y. and Fraser, E. (1989) "City size, economic development, and quality of life in China: New empirical evidence", *American Sociological Review* 54: 986–1003.

Bromley, D.W. (ed.) (1992) *Making the Commons Work: Theory, Practice and Policy.* San Francisco: ICS Press.

Brown, L. and Jacobson, J.L. (1987) *The Future of Urbanization: Facing the Ecological and Economic Constraints.* WorldWatch Paper No. 77, Washington, DC: WorldWatch Institute.

Chandler, D. (2003) "Expanding the research agenda of human rights: reply to Bellamy", *International Journal of Human Rights*, 7 (1): 128–140.

Cosgrove, W.J. and Rijsberman, F.R. (2000) *World Water Vision: Making Water Everybody's Business*. London: Earthscan.

Costanza, R., Andrade, F., Antunes, P., ven den Belt, M., Boersma, D., Boesch, D.F., Catarino, F., Hanna, S., Limburg, K., Low, B., Pereira, J.G., Rayner, S., Santos, R., Wilson, J. and Young, M. (1998) "Principles for sustainable governance of the oceans", *Science* 281: 198–199.

Costanza, R., Fisher, B., Ali, S., Beer, C., Bond, L., Boumans, R., Danigelis, N.L., Dickinson, J., Elliott, C., Farley, J., Gayer, D.E., MacDonald Glenn, L., Hudspeth, T., Mahoney, D., McCahill, L., McIntosh, B., Reed, B., Rizvi, S.A.T., Rizzon, D.M., Simpatico, T. and Snapp, R. (2007) "Quality of life: an approach integrating opportunities, human needs, and subjective well-being", *Ecological Economics* 61: 267–276.

Craig, D.J. (2008) "The deep sleep: are Americans waking up to global warming?", *Columbia Magazine*, Fall: 14–21. www.columbia.edu/cu/alumni/Magazine/Fall2008/feature1.html.

Cronon, W. (1991) *Nature's Metropolis: Chicago and the Great West.* 1st edn. New York: W.W. Norton.

Cronon, W. (1996) "The trouble with wilderness, or, getting back to the wrong nature", *Environmental History* 1 (1): 7–55. www.williamcronon.net/writing/Trouble_with_Wilderness_Main.html.

Daly, H.E. (1997) *Beyond Growth: The Economics of Sustainable Development.* Boston: Beacon Press.

Daly, H.E. (2005) "Economics in a full world", *Scientific American* 293 (3): 100–107.

Daly, H.E. and Farley, J. (2003) *Ecological Economics: Principles and Applications*. Washington, DC: Island Press.

Davis, G. (1998) *Creating Scenarios for Your Company's Future.* The 1998 Conference on Corporate Environmental, Health and Safety Excellence Bringing Sustainable Development Down to Earth, 28 April, New York City.

Diamond, J. (2005) *Collapse: How Societies Choose to Fail or Succeed.* New York: Viking Press.

Diener, E. and Biswas-Diener, R. (2002) "Will money increase subjective well-being? A literature review and guide to needed research", *Social Indicators Research* 57: 119–169.

Easterlin, R. (1974) "Does economic growth improve the human lot? Some empirical evidence". In: *Nations and Households in Economic Growth: Essays in Honor of Moses Abramovitz*. David, P.A. and Reder, M.W. (eds). New York: Academic Press.

Falkenmark, M. (2003) "Freshwater as shared between society and ecosystems: from divided approaches to integrated challenges", *Philosophical Transactions of the Royal Society B* 358: 2037–2049.

FAO (2008) *The State of Food Insecurity in the World 2008.* Rome: FAO.

Farley, J. (2008) "The role of prices in conserving critical natural capital", *Conservation Biology* 22 (6): 1399–1408.

Frank, R.H. (2007) *Falling Behind: How Rising Inequality Harms the Middle Class.* Berkeley: University of California Press.

Gallopin, G., Hammond, A., Raskin, P. and Swart, R. (1997) *Branch Points: Global Scenarios and Human Choice – A Resource Paper of the Global Scenarios Group*. Stockholm: Stockholm Environment Institute.

Gao, K. and McKinley, K.R. (1994) "Use of macroalgae for marine biomass production and CO_2 remediation: a review", *Journal of Applied Phycology* 6: 45–60.

Gore, A. (2008) "The climate for change", *New York Times*, 9 November.

Hall, D.O. and Overend, R.P. (eds) (1987) *Biomass: Regenerable Energy*. Chichester: Wiley and Sons.

Hansen, J. (2008) Letter to Michelle and Barack Obama, 29 December. www.columbia.edu/~jeh1/mailings/20081229_DearMichelleAndBarack.pdf.

Hardin, G. (1968) "The tragedy of the commons", *Science* 162 (3859): 1243–1248.

Hardin, G. (1998) "Extensions of 'The tragedy of the commons'", *Science* 280 (5364): 682–683.

Hegmon, M., Nelson, M.C. and Ruth, S.M. (1998) "Abandonment and reorganization in the Mimbres region of the American Southwest", *American Anthropologist* 100 (1): 148.

Hegmon, M., Peeples, M., Kinzig, A., Kulow, S., Meegan, C.M. and Nelson, M.C. (2008) "Social transformation and its human costs in the prehistoric U.S. Southwest", *American Anthropologist* 110: 313–324.

Himmel, M.E., Ding, S.-Y., Johnson, D.K., Adney, W.S., Nimlos, M.R., Brady, J.W. and Foust, T.D. (2007) "Biomass recalcitrance: engineering plants and enzymes for biofuels production", *Science* 315 (5813): 804–807.

Hobbs, R.J., Arico, S., Aronson, J., Baron, J.S., Bridgewater, P., Cramer, V.A., Epstein, P.R., Ewel, J.J., Klink, C.A., Lugo, A.E., Norton, D., Ojima, D., Richardson, D.M., Sanderson, E.W., Valladares, F., Vila, M., Zamora, R. and Zobel, M. (2006) "Novel ecosystems: theoretical and management aspects of the new ecological world", *Global Ecology and Biogeography* 15: 1–7.

Hoffert, M.I., Caldeira, K., Jain, A.K., Haites, E.F., Harvey, L.D.D., Potter, S.D., Schlesinger, M.E., Schneider, S.H., Watts, R.G., Wigley, T.M.L. and Wuebbles, D.J. (1998) "Energy implications of future stabilization of atmospheric CO_2 content", *Nature* 395: 881–884.

IPCC (2007) *Climate Change 2007: The Physical Science Basis.* Contribution of Working Group I to the Fourth Assessment Report of the Intergovernmental Panel on Climate Change. Solomon, S., Qin, D., Manning, M., Chen, Z., Marquis, M., Averyt, K.B., Tignor, M. and Miller, H.L. (eds). Cambridge: Cambridge University Press.

Jacobs, J. (1969) *The Economy of Cities.* New York: Random House.

Jäger, J. (2009) *Our Planet: How Much More can Earth Take*? London: Haus Publishing.

Jäger, J., Rothman, D., Anastasi, C., Kartha, S. and van Notten, P. (2008) "Training Module 6: Scenario development and analysis". In: *GEO Resource Book: A Training Manual on Integrated Environmental Assessment and Reporting*. www.unep.org/ieacp/iea/training/manual/module6.aspx.

Kahane, A., with Roux, P. (1992) "The Mont Fleur scenarios". *GBN Deeper News.* www.gbn.com.

Kahane, A. (1998) "Destino Colombia: a scenario-planning process for the new millennium". *GBN Deeper News.* www.gbn.com

Kahn, H. and Wiener, A.J. (1967) *The Year 2000: A Framework for Speculation on the Next Thirty-three Years*. New York: Macmillan.

Lane, R.E. (2000). *The Loss of Happiness in Market Democracies*. New Haven: Yale University Press.

Layard, R. (2005) *Happiness: Lessons from a New Science*. New York: Penguin.

Lewis, N.S. and Nocera, D.G. (2006) "Powering the planet: chemical challenges in

solar energy utilization", *Proceedings of the National Academy of Sciences, USA* 103 (43): 15729–15735.
Linden, E. (1996) "The exploding cities of the developing world", *Foreign Affairs* 75 (1): 52–65.
McGranahan, G., Marcotullio, P.J., Bai, X., Balk, D., Braga, T., Douglas, I., Elmqvist, T., Rees, W., Satterthwaite, D., Songsore, J., Zlotnik, H., Eades, J. and Ezcurra, E. (2005) "Urban systems". In: *Current State and Trends: Findings of the Condition and Trends Working Group – Ecosystems and Human Well-being*, Vol. 1. Washington, DC: Island Press.
McKibben, B. (1995) *Hope, Human and* Wild: *True Stories of Living Lightly on the Earth*. Minneapolis: Milkweed Editions.
McKinney, M.L. (2008) "Effects of urbanization on species richness: a review of plants and animals", *Urban Ecosystems* 11 (2): 161–176.
Marcotullio, P.J. and Schulz, N.B. (2007) "Comparison of energy transitions in the United States and developing and industrializing economies", *World Development* 35 (10): 1650–1683.
Martinelli, L.A. and Filoso, S. (2008) "Expansion of sugarcane ethanol production in Brazil: environmental and social challenges", *Ecological Applications* 18 (4): 885–898.
Max-Neef, M.A. (1991) *Human Scale Development: Conception, Application and Further Reflections*. New York: Apex Press.
Meadows, D.H. (1996) "Envisioning a sustainable world". In: *Getting Down to Earth, Practical Applications of Ecological Economics*, Robert Costanza, Olman Segura and Juan Martinez-Alier (eds). Washington, DC: Island Press.
Meadows, D. (1999) *Leverage Points: Places to Intervene in a System*. The Sustainability Institute. www.sustainabilityinstitute.org/pubs/Leverage_Points.pdf.
Moriarty, P., Batchelor, C. and Laban, P. (2005) "Using visions, scenarios and strategies within the EMPOWERS planning cycle for IWRM". EMPOWERS Working Paper No. 4. www.negowat.org/curso/Modulo%20II/Documentos/Scenario%20planning.pdf.
Myers, N. and Kent, J. (2001) *Perverse Subsidies: How Tax Dollars can Undercut the Environment and the Economy*. Washington, DC: Island Press.
Nakicenovic, N. and Swart, R. (eds) (2000) *Emissions Scenarios: Special Report of the Intergovernmental Panel on Climate Change*. Cambridge: Cambridge University Press.
Nordhaus, W. (2007) "Alternative measures of output in global economic–environmental models: purchasing power parity or market exchange rates?", *Energy Economics* 29 (3): 349–372.
Norton, B.G. (2005) *Sustainability: A Philosophy of Adaptive Ecosystem Management*. Chicago: University Of Chicago Press.
Olsson, P., Gunderson, L.H., Carpenter, S.R., Ryan, P., Lebel, L., Folke, C. and Holling, C.S. (2006) "Shooting the rapids: navigating transitions to adaptive governance of social-ecological systems", *Ecology and Society* 11 (1): 18. www.ecologyandsociety.org/vol. 11/iss1/art18.
Ostrom, E. (1990) *Governing the Commons: The Evolution of Institutions for Collective Action*. Cambridge: Cambridge University Press.

Ostrom, E., Burger, J., Norgaard, R.B., Policansky, D. and Goldstein, B.D. (eds) (2001) *Protecting the Commons: A Framework for Resource Management in the Americas.* Washington, DC: Island Press.

PBL (2009) "Getting into the right lane for 2050". Netherlands Environmental Assessment Agency. www.pbl.nl/en/publications/2009/Getting-into-the-Right-Lane-for-2050.html.

Peterson, G.D., Beard, D., Beisner, B., Bennett, E., Carpenter, S., Cumming, G., Dent, L. and Havlicek, T. (2003) "Assessing future ecosystem services: a case study of the northern highland lake district, Wisconsin", *Conservation Ecology* 7 (3): 1. www.consecol.org/vol. 7/iss3/art1.

Pew Center (2008) *A Deeper Partisan Divide over Global Warming.* Survey Report, May 2008. The Pew Research Center for the People and the Press. http://people-press.org/report/417/a-deeper-partisan-divide-over-global-warming.

Pew Center (2009) *Modest Support for "Cap and Trade" Policy: Fewer Americans see Solid Evidence of Global Warming.* Survey Report, October 2009. The Pew Research Center for the People and the Press. http://people-press.org/reports/pdf/556.pdf.

Ponting, C. (2007) *A New Green History of the World: The Environment and the Collapse of Great Civilizations*. London: Vintage.

Prugh, T., Costanza, R. and Daly, H. (1999) *The Local Politics of Global Sustainability*. Washington, DC: Island Press.

Ramanathan, V. and Feng, Y. (2008) "On avoiding dangerous anthropogenic interference with the climate system: formidable challenges ahead", *Proceedings of the National Academy of Sciences USA* 105 (38): 14245–14250.

Raskin, P., Banuri, T., Gallopin, G., Gutman, P., Hammond, A., Kates, R. and Swart, R. (2002) *Great Transition: The Promise and Lure of the Times Ahead.* Boston: Stockholm Environment Institute.

Raupach, M., Marland, G., Ciais, P., le Quere, C., Canadell, J.G., Klepper, G. and Field, C.B. (2007) "Global and regional drivers of accelerating CO_2 emissions", *Proceedings of the National Academy of Sciences USA* 104: 9913–9914.

Rotmans, J., Kemp, R., van Asselt, M., Geels, F., Verbong, G. and Molendijk, K. (2000) *Transitions and Transition Management.* International Centre for Integrative Studies. www.icis.unimaas.nl/projects/nmp4/downs/00_35_ab.pdf.

Sagar, A.D. and Kartha, S. (2007) "Bioenergy and sustainable development?", *Annual Review of Environment and Resources* 32: 131–167.

Satterthwaite, D. (2007) *The Transition to a Predominantly Urban World and its Underpinnings.* IIED Human Settlements Discussion Paper Series No. 91. www.iied.org/pubs/pdfs/10550IIED.pdf.

Schwartz, P. (1996) *The Art of the Long View*. New York: Currency Doubleday.

Searchinger, T., Heimlich, R., Houghton, R.A., Dong, F., Elobeid, A., Fabiosa, J., Tokgoz, S., Hayes, D. and Yu, T.-H. (2008) "Use of US croplands for biofuels increases greenhouse gases through emissions from land-use change", *Science* 319 (5867): 1238–1240.

Shellenberger, M. and Nordhaus, T. (2009) *Break Through: Why We Can't Leave Saving the Planet to Environmentalists*. Boston: Mariner Books.

SIWI (2005a) *Towards Hydrosolidarity: Ample Opportunities for Human Ingenuity.* Stockholm International Water Institute Report. www.siwi.org/documents/Resources/Reports/Report_Towards_hydrosolidarity_2005.pdf.

SIWI (2005b) *Making Water a Part of Economic Development: The Economic Benefits of Improved Water Management and Services.* Stockholm International Water Institute Report. www.siwi.org/documents/Resources/Reports/CSD_Making_water_part_of_economic_development_2005.pdf.

SIWI (2007) *On the Verge of a New Water Scarcity: A Call for Good Governance and Human Ingenuity.* Stockholm International Water Institute Policy Brief. www.siwi.org/documents/Resources/Papers/Paper11_Water_Scarcity_2007.pdf.

Smart, A. and Smart, J. (2003) "Urbanization and the global perspective", *Annual Review of Anthropology* 32: 263–285.

Smil, V. (2003) *Energy at the Crossroads: Global Perspectives and Uncertainties.* Cambridge, MA: MIT Press.

Sondeijker, S. (2009) *Imagining Sustainability.* Rotterdam: Erasmus Universiteit.

Steffen, W., Sanderson, A., Tyson, P.D., Jäger, J., Matson, P.M., Moore III, B., Oldfield, F., Richardson, K., Schellnhuber, H.J., Turner II, B.L. and Wasson, R.J. (2004) *Global Change and the Earth System: A Planet Under Pressure.* New York: Springer.

Stern, N. (2006) *The Stern Review on the Economics of Climate Change.* UK Government Cabinet Office and HM Treasury. www.hm-treasury.gov.uk/sternreview_index.htm.

Toffler, A. (1984) "Foreword". In: *Order Out of Chaos: Man's New Dialogue with Nature.* Prigogine, I. and Stengers, I. London: Flamingo.

UNDP (2008) *Human Development Report 2007/2008: Fighting Climate Change – Human Solidarity in a Divided World.* http://hdr.undp.org/en/reports/global/hdr2007-2008.

UNEP (2000) *Earth and Faith: A Book of Reflection for Action.* Bassett, L., Brinkman, J.T. and Pedersen, K.P. (eds). New York: UNEP.

UNEP (2002) *Global Environment Outlook 3: Past, Present and Future Perspectives.* London: Earthscan.

UNEP (2007) *Global Environment Outlook 4.* UNEP. www.unep.org/geo/geo4/report/GEO-4_Report_Full_en.pdf.

Van der Heijden, K. (1996) *Scenarios: The Art of Strategic Conversation.* Chichester: John Wiley.

Van der Leeuw, S.E. (2005) "Climate, hydrology, land use, and environmental degradation in the lower Rhone Valley during the Roman Period", *Geoscience* 337 (1–2): 9–27.

Van Notten, P., Rotmans, J., van Asselt, M.B.A. and Rothman, D. (2003) "An updated scenario typology", *Futures*, 35: 423–445.

Wack, P. (1985) "Scenarios: shooting the rapids", *Harvard Business Review*, November–December: 139–150.

Weaver, P.M. and Rotmans, J. (2006) "Integrated sustainability assessment: what is it, why do it, and how?" *International Journal of Innovation and Sustainable Development* 1(4): 284–303.

World Bank (1999) *World Development Report 1999–2000: Entering the 21st Century.* Oxford: Oxford University Press.

Zarrilli, S. and Burnett, J. (2008) *Making Certification Work for Sustainable Development: The Case of Biofuels.* United Nations Conference on Trade and Development. www.unctad.org/templates/Download.asp?docid=10481&lang=1&intItemID=3830.

Index